Chemistry at the races

The work of the Horseracing Forensic Laboratory

Compiled by Ted Lister

Chemistry at the races – the work of the Horseracing Forensic Laboratory

Compiled by Ted Lister

Edited by Colin Osborne and Maria Pack

Designed by Imogen Bertin

Published and distributed by Royal Society of Chemistry

Printed by Royal Society of Chemistry

For further information on other educational activities undertaken by the Royal Society of Chemistry write to:

Education Department
Royal Society of Chemistry
Burlington House
Piccadilly
London W1J 0BA

Information on other Royal Society of Chemistry activities can be found on its websites:
http://www.rsc.org
http://www.chemsoc.org
http://www.chemsoc.org/LearnNet contains resources for teachers and students from around the world.

ISBN 0-85404–385–3

British Library Cataloguing in Publication Data.

A catalogue for this book is available from the British Library.

Foreword

The chemical sciences pervade almost all aspects of modern life from the clothes we wear to the food we eat and the pastimes we enjoy. This resource shows how a combination of modern techniques are used to ensure that horseracing is both fair and prevents abuse of the horses involved. It is hoped teachers will use it to provide an up-to-date and interesting context for their work and so enthuse the next generation of scientists.

Professor Steven Ley CChem FRSC FRS
President, The Royal Society of Chemistry

RS•C

Contents

RS•C

Introduction

This material is based on the work of the Horseracing Forensic Laboratory (HFL) located near Newmarket in Suffolk. HFL screens samples of urine and sometimes blood from racehorses and other animals to detect and identify traces of prohibited substances.

The material is the result of a workshop held at HFL and organised by The Royal Society of Chemistry. A group of chemistry teachers spent a day at HFL and was given a presentation and tour of the laboratories. The following day was spent brainstorming and drafting the material, which is presented here in edited form.

The personnel involved were:

Ben Faust, Loughborough Grammar School;
Simon Hudson, HFL;
Ted Lister, Educational Consultant;
Steve Maynard, HFL;
Brian McVicar, Croydon High School for Girls; and
Colin Osborne, Royal Society of Chemistry.

Acknowledgements

The Royal Society of Chemistry (RSC) thanks all the participants, especially Simon Hudson, and Steve Maynard and their colleagues at HFL. The RSC would also like to thank Maura O'Connor, Veterinary Officer of the Irish Turf Club, and Caroline Norris, photographer, for their assistance in providing pictures of the sampling process. Agilent kindly supplied pictures of GCMS equipment.

The material

This includes teachers' notes (below) and five sections of student material that may be copied for use in class. The student material may be downloaded as .pdf files or as Word documents that may be further edited by the teacher if required. Each piece of student material consists of a passage with questions to fulfil the following functions:

- as summative comprehension questions, *ie* questions to test students' understanding after they have read the material;
- as formative comprehension questions *ie* to help students' understanding as they read the material; and
- as a way of highlighting that the chemical principles used by chemists in real life situations are the same as those studied at school or in college.

The teachers' notes give brief answers to the questions and in some cases extra detail that may be of use to the teacher. Teachers who do not wish to use the material in its entirety with their classes may find it useful both to refresh their own knowledge and as a source of examples to use in their teaching.

The student material is as follows:

Part 1. Chemistry at the races –
the work of the Horseracing Forensic Laboratory (post-16 level)
This material presents an overview of the work of HFL, explaining the reasons for drug

testing in racehorses, discussing the sampling procedure and dealing briefly with the two main analytical techniques – Gas Chromatography-Mass Spectrometry (GCMS) and immunoassay. This may be used as an introduction to parts 2– 4.

Part 2. Detecting drugs in horse urine by immunoassay (post-16 level)
This section focuses on immunoassay and the technique is discussed in more detail than in part 1.

Part 3. Drug metabolism (post-16 level)
This material discusses some of the changes that take place to drug molecules in the liver and points out that analytical techniques for drug detection may detect metabolite molecules as well as the original drug.

Part 4. Identifying drugs in horse urine by Gas Chromatography – Mass Spectrometry (GCMS) (post-16 level)
This section focuses on GCMS and the technique is discussed in more detail than in part 1.

Part 5. Testing for drugs in racehorses using chromatography (11–14 year olds)
The use of gas chromatography is briefly described and is compared and contrasted with paper chromatography.

RS•C

Teachers' notes and answers

Chemical nomenclature

This material contains references to many substances (particularly drugs) whose systematic names are long and complex. In general, these have been referred to by their trivial names except where students might be expected to recognise the systematic name, in which cases both trivial and systematic names are given (for example salicylic acid, 2-hydroxybenzoic acid).

Part 1 Chemistry at the races – the work of the Horseracing Forensic Laboratory

Answers

1. a) To preserve the second sample in as nearly as possible the same state as the first.

b) Chemical changes will occur in the sample, which may affect the concentrations of any drugs or their metabolites. Freezing prevents the motion of molecules from place to place, hence preventing collisions and thus stopping chemical reactions.

c) For many reactions, the rate is halved.

d) The reaction rate would drop to one quarter of its value at 20 °C.

e) In a frozen sample at 0 °C, molecular motion (from place to place) has almost stopped, so the rates of reaction will drop virtually to zero. In a liquid sample at 0 °C, molecular motion continues, so chemical reactions will continue to take place.

2. a) A sample containing the drug remains colourless while a sample with no drug turns yellow. The yellow-coloured solution will allow yellow light only to pass through, absorbing other colours.

b) Blue. Blue is absorbed by the yellow solution (see (a) above), so the greater its concentration, the more light will be absorbed. The solution allows yellow light to pass through, so increasing its concentration will have no effect on the amount of yellow light passing through.

3 a) One or more type(s) of intermolecular forces – van der Waals, dipole-dipole or hydrogen bonding.

b) Propan-1-ol would have the longest retention time and propane the shortest.

The –OH groups on the propan-1-ol molecule would be able to form strong hydrogen bonds with the –OH groups on the coating via the oxygen and the hydrogen atoms.

Propane would have the shortest retention time because it can form only weak van der Waals bonds with the coating.

RS•C

Propanone would have an intermediate retention time as it can form hydrogen bonds with the –OH groups of the coating through the oxygen of the carbonyl group only – none of its hydrogen atoms can form hydrogen bonds as they are all bonded to carbon atoms.

4 a) i) The electron beam strikes the sample molecule and ejects an electron from it, forming a positive ion.

(ii) Either
If the ejected electron comes from one of the shared pairs in one of the bonds, this will weaken the bond, which may then break.
Or
If the energy of the electron-sample molecule collision is greater than the bond energy of one of the bonds in the sample molecule, then this bond may break.

b) Positive ions will be formed. These will normally be singly charged, but occasionally doubly charged ions may form if an electron from the beam collides with an ion that has already formed.

5 a) i) The peak at mass 43 may correspond to CH_3CO^+.
The peak at mass 121 may correspond to $C_6H_4COOH^+$.
ii) The peak at mass 180 represents the molecular (parent) ion, *ie* the unfragmented sample molecule.

b) To reduce pain and inflammation and allow it to run despite an injury.

Part 2 Detecting drugs in horse urine by immunoassay

Notes

Most drug molecules are too small to elicit an immune response on their own. So, to raise antibodies to, say, morphine, the morphine molecule has to be conjugated to a larger (protein) molecule. This technicality has been ignored in the students' material, but some teachers may wish to point this out to some students.

The three controls used in the immunoassay technique are as follows.

1. A negative blank which should, of course, show no drug present.
2. A 'low control'. This is a sample made up of horse urine (or possibly artificial urine) to which has been added a known, low, concentration of the drug being sought. This acts as a check on the accuracy of the analysis at low concentrations.
3. A 'high control'. This is a sample made up of horse urine (or possibly artificial urine) to which has been added a known, high, concentration of the drug being sought. This acts as a check on the accuracy of the analysis at high concentrations.

Answers

1. a) If the antibody were added first, all of it would bind to the drug molecules attached to the walls of the well. As this binding is essentially irreversible, no antibody would remain to bind to the free drug in the positive sample.

b) The antibody will only bind to its own antigen (the drug molecule in question).

2. a) The nitrogen atom has a lone pair of electrons that allows diethanolamine to accept a proton (H^+ ion) and thus act as a base.
$(HOC_2H_4)_2NH + H^+ \rightarrow (HOC_2H_4)_2NH_2^+$

b) A buffer is a solution that resists change of pH when small amounts of acid or base are added.

RS•C

c) The conjugate acid of diethanolamine must be added, *ie* a compound containing the ion $(HOC_2H_4)_2NH_2^+$, *eg* $(HOC_2H_4)_2NH_2^+Cl^-$.

d) A buffer is required to fix the pH of the solution at the optimum for this particular enzyme.

e) Hydrolysis.

3. The yellow-coloured solution will allow yellow light to pass through, absorbing other colours including blue. So the greater the concentration of the yellow solution, the more blue light will be absorbed. The solution allows all the yellow light to pass through, so increasing its concentration will have no effect on the amount of yellow light passing through.

4. See Figure 1.

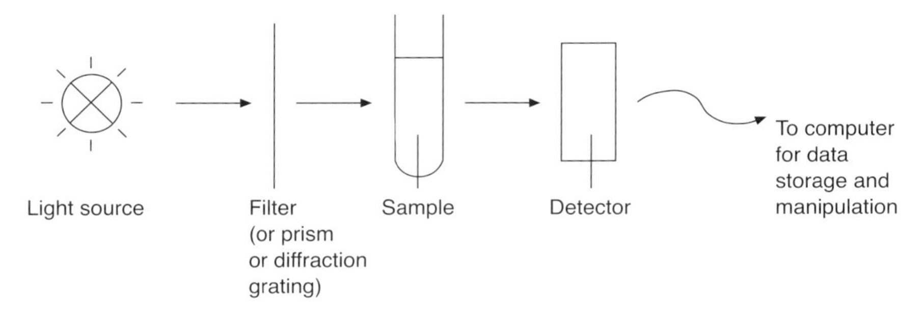

Figure 1

The essential parts are:

- a source of light;
- a filter (or prism or diffraction grating) to select a suitable wavelength;
- a suitable container for the sample; and
- a detector.

Students might also draw a computer to store and manipulate the data.

Part 3 Drug metabolism

Notes

The half-life of acepromazine in horses is approximately 24 hours.

The cytochrome oxidase P 450 enzymes are so called because they absorb light at a wavelength of 450 nm.

Answers

1. a) The half lives are the same; 1 hour.

   b) It has dropped to 1/16 of its original value.

2. Both have six-membered rings containing an oxygen atom with –OH functional groups. Glucuronic acid has a carboxylic acid functional group *ie* the CH_2OH group has been oxidised to –COOH.

3. a) See Figure 2.

Figure 2

b) Both angles are approximately 60°.

c) These angles are normally approximately 109.5°.

d) This suggests that epoxides will be unstable due to ring strain.

4. The phenol group is more water-soluble than a benzene ring because the –OH group can hydrogen bond to water.
An alcohol is more soluble in water than a carbonyl group because the –OH group can form more hydrogen bonds with water than can the C=O group of the carbonyl.
N–H and O–H are able to form more hydrogen bonds with water than can N-R and O–R so the de-alkylated derivatives will be more water-soluble.

5. The metabolites are more soluble in water than the original molecules and can be more easily excreted in urine.

6. a is reduction; b is oxidation; c is oxidation; d is reduction.

7. The N-CH_3 groups could be de-alkylated.

8. a) It will make it more soluble because of the many –OH groups on the glucuronic acid group. These can form hydrogen bonds with water.

b) The glucuronide can be more easily excreted in urine.

9. Any suitable ones *eg* sugars, haemoglobin and inorganic salts (in blood), urea and inorganic salts (in urine). Water, of course is present in both.

10. The de-alkylation of the –O–CH_3 group.

Part 4 Identifying drugs in horse urine by Gas Chromatography–Mass Spectrometry (GCMS)

Notes

The enzymes used in sample preparation remove, for example, glucuronic acid residues that have been added in phase II metabolic reactions.

Answers

1. A buffer is a solution that resists change in its pH when small amounts of acid or base are added. It is used to maintain the pH at the optimum pH of the enzyme.

2. a) The –COOH group.

b) The tertiary amine groups.

3. Intermolecular forces such as van der Waals forces, dipole-dipole forces or hydrogen bonding, depending on the structures of the molecules involved.

4. The coating of the column corresponds to the paper, the helium corresponds to the water and the retention time corresponds to the distance moved along the paper.

5. The type of coating on the column, the flow rate of the helium, the temperature.

6. If the inside of the mass spectrometer were not kept under a high vacuum, charged fragments would collide with air molecules as they flew through the instrument.

7. All the main peaks (such as those at m/e 42, 57, 71, 91, 103, 172, 218 and 247) are present in both spectra, although their heights are slightly different. These spectra almost certainly represent a match, *ie* pethidine was present in the urine sample.

Part 5 Testing for drugs in racehorses using chromatography

Notes

The procedure described for the initial cleaning up of the urine sample by pouring it through a column with a tap is not exactly as done at HFL. HFL use a cartridge containing silica rather than a column with a tap at the base. The column method is described because the principle is the same, and it was felt that this apparatus would be more familiar to students of this age.

It was not thought appropriate to give any detail of mass spectrometry in material for students in the 11–14 age group. More detail for the teacher's use can be found in the chapters *Chemistry at the races – the work of the Horseracing Forensic Laboratory* and *Identifying drugs in horse urine by Gas Chromatography – Mass Spectrometry.*

The gas chromatograms shown in Figures 5 and 6 are simplified. One typical of those obtained from horse urine can be found in *Identifying drugs in horse urine by Gas Chromatography – Mass Spectrometry*, Figure 4.

Answers

1. Examples include: witnessing of the sampling procedure, sealing the samples in containers that make any tampering obvious, transport by courier.

2. So that the second sample could be rechecked or independently tested.

3. **a)** **i)** silica.

   ii) solution or liquid of pH 4 (do not accept water).

   b) Test with pH meter/universal indicator solution or paper.

4. **a)** It could burn/combust the substance.

   b) Any unreactive gas *eg* helium, nitrogen.

5. The peak at time 22 minutes represents a drug. This peak is found in the sample from the horse and the reference sample only – all the other peaks were in either the system blank, the biological blank or both.

6. Take urine samples from horses that have been treated by a vet with the drug. Compare the urine samples taken after treatment with samples taken before. Identify any new substances and then look at their chemical structures to try to show that they could be derived from the drug. Other sensible suggestions should be accepted.

7. Eventing (or other equestrian sport), greyhound racing *etc.*

8. Look for a discussion of ethical issues – horses cannot decide of their own free will to take drugs or not.

RS•C

Chemistry at the races

The work of the Horseracing Forensic Laboratory

Drug testing of animals

Aliysa fails Oaks drug te

'Stopping drug' fears

They were doped!

Bravefoot and Norwich tested positive after Doncaster flops

'I don't know what's going on. My horse had something it shouldn't have done, but from where that came I haven't a clue'

JOSH GIFFORD

PRICES & OFFERS

THE BRENT WALKER

Figure 1 'Doping' of horses can hit the headlines as shown in these articles from the Racing Post, reproduced with permission.

Most people will know that sportsmen and women are regularly tested for drug use both in competition and in training. Drugs are also banned in sporting competitions involving animals, including horses and dogs. Cases where household names fail drug tests hit the headlines, Figure 1. The authorities that administer these sports, for example The Jockey Club of Great Britain for horseracing, have strict rules about drug use. This article concentrates on horseracing, but the principles involved are similar for all such authorities. The Horseracing Forensic Laboratory (HFL), situated near Newmarket, in the heart of racing country, carries out drug testing on horses (and on dogs) competing in the UK and abroad.

There are several reasons for drug testing of animals in competition, these include:

1. To ensure that outcomes of races are not manipulated illegally (for betting purposes, for example).
2. To ensure that the welfare of the animals is not put at risk. For example an injured animal might be given pain-killers to allow it to compete when it is not fit to do so, thereby worsening the injury. Humans may choose whether or not to use a drug in this situation, animals cannot.
3. Stud and animal sales issues. 'Stud' refers to the breeding of racehorses to sell the offspring. Someone buying a valuable horse (or its semen) will need to know that its racing performances have been genuine and not enhanced by the use of drugs.

The rules relating to drug use in a particular sport are made by that sport's governing body. HFL in Newmarket analyses samples of urine (or, occasionally of blood) and reports back to the governing body. Each sample arriving at HFL is coded so that the analysts do not know the identity of the animal concerned, or even the sporting event it has come from. This is important to ensure that there can be no accusations of bias. Any report of a positive test is passed on to the appropriate governing body, which decides whether any further action should be taken. In this way the laboratory can maintain its professional integrity – it is completely independent and impartial.

The rules governing what drugs can be given to horses differ in principle from those governing human athletes. For human athletes, there is a list of banned categories of substances. For horses, the principle is that any drug that is physiologically active is banned. You can find more detail about classes of drugs that are banned in horses and the body systems that they act upon at:

http://www.thejockeyclub.co.uk/jockeyclub/html/racing/substances.htm (accessed October 2001).

Sampling

It is not practical to test every racehorse. At a race meeting it is the stewards in charge of the event who decide which horses should be tested. There is no predictable formula for deciding which horses to test because this could help potential dopers to evade detection. However, winners, and horses that have run much better or much worse than expected, are liable to be selected. In this way the mere risk of a dope test being done has a deterrent effect.

Blood samples can be taken for analysis, but they give only a 'snapshot' of what is in the animal's system at the actual time of sampling – as the blood circulates through the liver, especially, changes will take place to the chemicals in it – see *Drug Metabolism*. More usually, urine is sampled for analysis. Unlike blood, urine's composition does not constantly vary – once blood has passed through the kidneys and excretion products have been removed into the urine, those products no longer change, they are merely stored in the bladder. Consequently, urine analysis gives information on what has been in the body as well as what is there at the time of sampling.

Drugs can remain in the body for several weeks and can also be identified from their metabolites, the substances that they are converted into in the liver. This means that illegal substances can be detected a considerable time after they have been given.

The sensitivity of modern analysis requires trainers to know exactly what has been fed to a horse. This could even include chocolate fed to it by well-meaning strangers; chocolate contains the stimulant theobromine. Security measures are taken to ensure that strangers cannot gain access to horses, so stable lads might be required to sleep in the stables the night before a race.

Keeping samples secure

After a race, the horses to be sampled are escorted to the doping unit, where a specimen must be collected. This might involve a wait, but rustling straw has been known to encourage horses to urinate. The sample is collected in a plastic bag held in a net on the end of a telescopic pole (rather like a fishing net). The urine is then poured into two identical 250 cm^3 polythene bottles. The bottles are then sealed by a vet, packed with a cooling pack like those used in picnic boxes and marked with a barcode before being sent to HFL for analysis.

It is essential that the samples analysed are not tampered with on their way to HFL. There are a number of tamper-evident layers of packaging. These include:

- The address label of the sample box covers the closure, so will be obvious if the box has been opened.

RS•C

- The sample bottles are sealed in a polythene envelope. If the flap is lifted, the word 'void' shows, and the heat-sealed sides of the envelope have lettering in the polymer so that it is obvious if an attempt has been made to slit the sides to get to the bottles.
- The sample bottles have tamper-evident plastic caps with lugs that must be broken to open them (some drinks sold in supermarkets have this type of cap).
- Once this cap has been removed, a heat-sealed diaphragm must be cut through to allow the contents to be poured out. This seal is applied by the vet, and is so strong that neither the bottle nor the seal breaks even if the bottle is stamped on.

The materials required for taking and packaging samples are sent to the racecourse in the form of a kit, Figure 2, timed to arrive just before the meeting to eliminate the potential for them to be interfered with.

Figure 2 Sampling kit

If any of the seals have been tampered with, the lab reports the details to the regulatory authority which then decides whether or not to have the sample analysed.

Analysis of a sample

When the lab receives a sample, one of the two bottles is put in a freezer to allow later analysis, or analysis by another independent lab if the first findings are challenged. The bottle that is to be used is opened, and its contents are poured out – no apparatus such as a pipette is inserted because this could introduce contamination. Two samples are taken, one is analysed by a method called immunoassay, and the other by Gas Chromatography-Mass Spectrometry (GCMS).

Q1. Part of the athlete Diane Modahl's defence against an allegation of drug taking was that her sample had been allowed to stand for some time in a hot laboratory before testing.

a) Why is freezing of the second sample necessary to allow it to be tested later?

b) Explain what might happen if the sample were not frozen and how freezing prevents this.

c) As a rule of thumb, what happens to the rate of a chemical reaction when it is cooled by 10 °C?

d)	**A typical laboratory might be at a temperature of 20 °C. What would happen to reaction rates if a sample were cooled to 0 °C but remained liquid?**
e)	**Why would freezing the sample so that it became solid at 0 °C be more effective than cooling the sample to a liquid at the same temperature?**

Immunoassay

In this method, the neat sample is treated with a solution containing antibodies. An antibody is a protein molecule produced by the body that will bind very specifically to another molecule called its antigen. Certain antibodies can be produced that bind specifically to particular drug molecules. Treatment of the sample with antibodies followed by enzymes and other reagents brings about a colour change so that a sample containing the drug targeted by a particular antibody stays colourless, and a sample without it goes yellow. The amount of colour can be read by a colorimeter, a device that shines a beam of light through the sample. More light will pass through a sample containing the drug than one without it. About 30 drugs are tested for by this method. These are ones that, for various reasons, are not easily measured by GCMS or ones present in low levels, for which immunoassay is particularly sensitive.

Q2. a)	**When a colorimeter is used in the immunoassay test, explain why 'more light will pass through a sample containing the drug than one without it'.**
(b)	**Which colour of light would be most suitable to measure the depth of colour of the yellow solution in the immunoassay test – yellow or blue? Explain your answer.**

Gas chromatography-mass spectrometry (GCMS)

Before this test, the urine sample is passed through a cartridge containing about a 10 mm depth of specially treated silica particles. These absorb the groups of compounds likely to contain drugs and their breakdown products, and let through others compounds such as inorganic salts so that they do not interfere with the analysis. The cartridge is then washed with solvents that remove more of the urine components whilst leaving any drugs in the silica. This is in effect a cleaning step. The mixture containing potential drugs is then washed from the silica with solvents of different pHs and separated by gas chromatography (GC).

Here the mixture is injected onto a chromatography column. This is a 25 m-long silica tube with an internal diameter of just 0.25 mm. (This means that it is flexible enough to be coiled into a spiral about 15 cm in diameter and placed in a temperature controlled oven, Figure 3.) The inside of the tube is specially coated and a stream of helium gas flows through the tube at about 1–2 $cm^3 \ min^{-1}$. This carries the components of the mixture through the tube. Some, however, move more slowly than others, as they tend to 'stick' to the coating of the tube. The 'stickier' they are, the more they tend to bond to the coating and the longer they take to come out of the tube. The time taken for a component to travel the length of the column is called its retention time. Typical retention times may be several minutes.

RS•C

Figure 3 Agilent 5973N GCMSD instrument, showing column inside oven

Q3. **a)** Suggest what type of bonding occurs between the molecules in a mixture and the coating of the GC tube to make them 'stick' to it.

b) Imagine that one type of GC tube coating has many –OH groups. Suggest which of the following molecules would have (i) the longest (ii) the shortest retention time: propan-1-ol, propane, propanone. Explain your answers.

Under a specified set of conditions, any component of the mixture will always have the same retention time. So its retention time can be used to help identify it. This is the gas chromatography (GC) part of GCMS. As each component of the mixture leaves the column, a detector measures its amount. The information is stored in a computer and can be presented as a graph of amount of substance against time. This is called a chromatogram. A typical one is shown in Figure 4. Each peak corresponds to a different component of the mixture. The taller the peak (strictly speaking the greater its area), the more there is of that component. In Figure 4, the numbers printed by some of the peaks are the retention times of those peaks. The computer used to store, manipulate and print the data has been used to mark the retention times of main peaks of the chromatogram to make it easier to interpret.

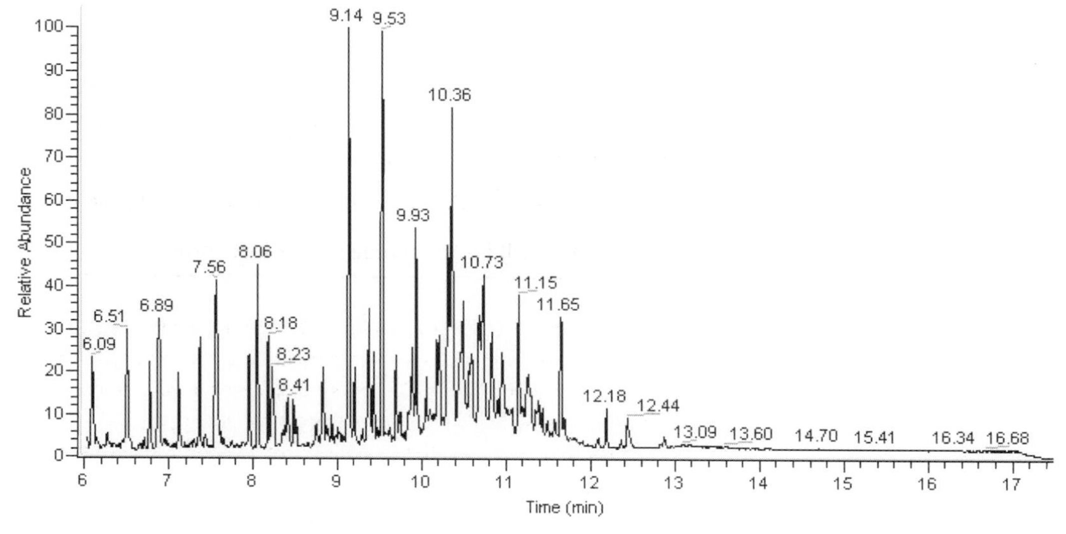

Figure 4 A typical gas chromatogram

RS•C

However, finding a peak in the chromatogram with the same retention time as a known drug is not enough to be certain that that drug is present. Two quite different substances might coincidentally have the same retention time. Here is where the mass spectrometer (MS) part of GCMS comes in. As the components leave the column, they are directed into a mass spectrometer. This fires a beam of electrons at the molecule, which becomes ionised and breaks up into fragments. The instrument separates these fragments by their masses, stores the data, and plots a graph of the relative amount of each fragment against mass, see Figure 5. This is called a mass spectrum.

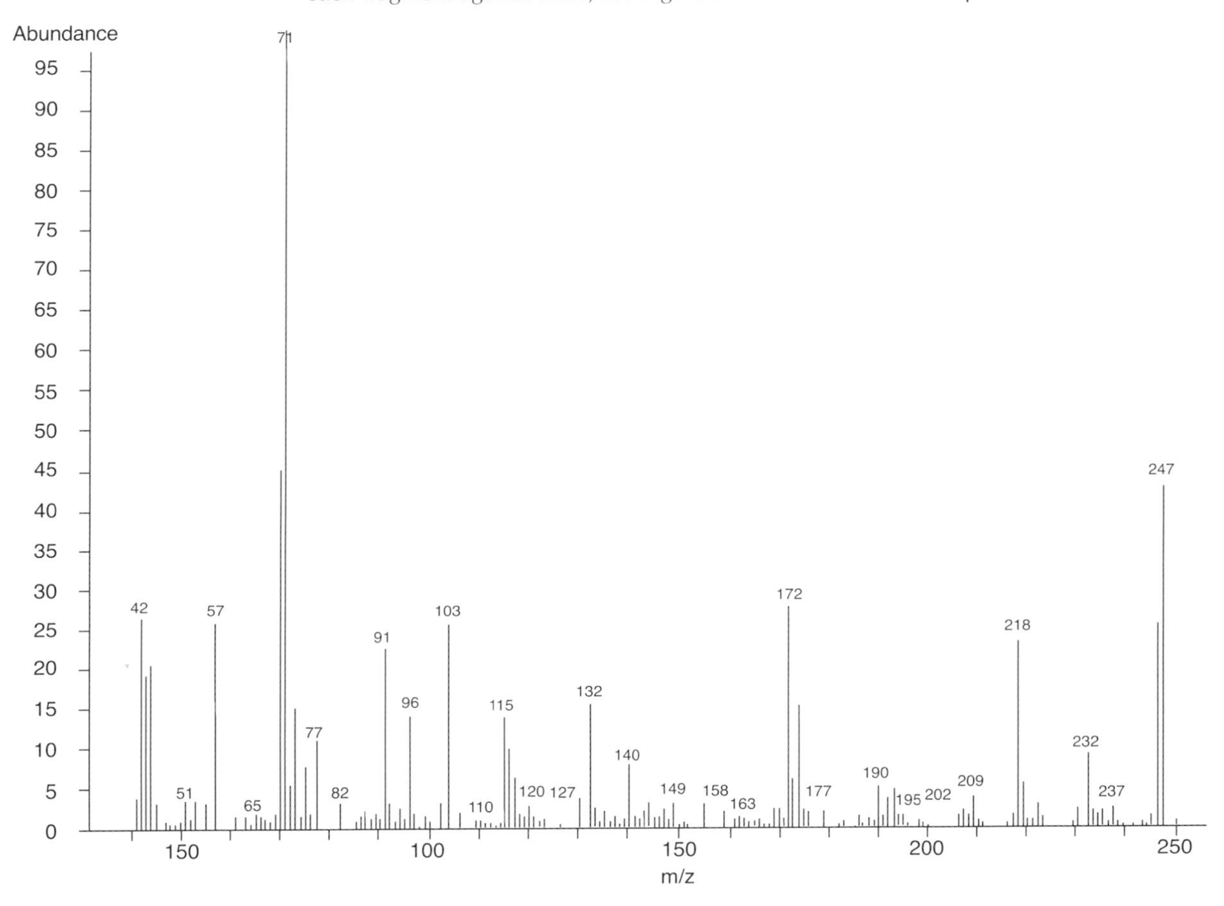

Figure 5 A typical mass spectrum

Q4. In the mass spectrometer, a beam of electrons is used to (i) ionise and (ii) fragment the sample molecules.

a) Explain how the electron beam brings about each of these processes.

b) What will be the charge on the ions formed?

Each substance has its own individual pattern of fragments, so a substance can be identified by comparing its mass spectrum with a mass spectrum known to be of that substance. The computer can easily match a mass spectrum obtained from GCMS with a database of many thousands of previously analysed spectra. A drug will be identified if both the mass spectrum and the retention time match those of a known standard.

In fact, the combination of chromatogram and mass spectrum for each of its peaks generates a three-dimensional data set as shown in Figure 6.

RS•C

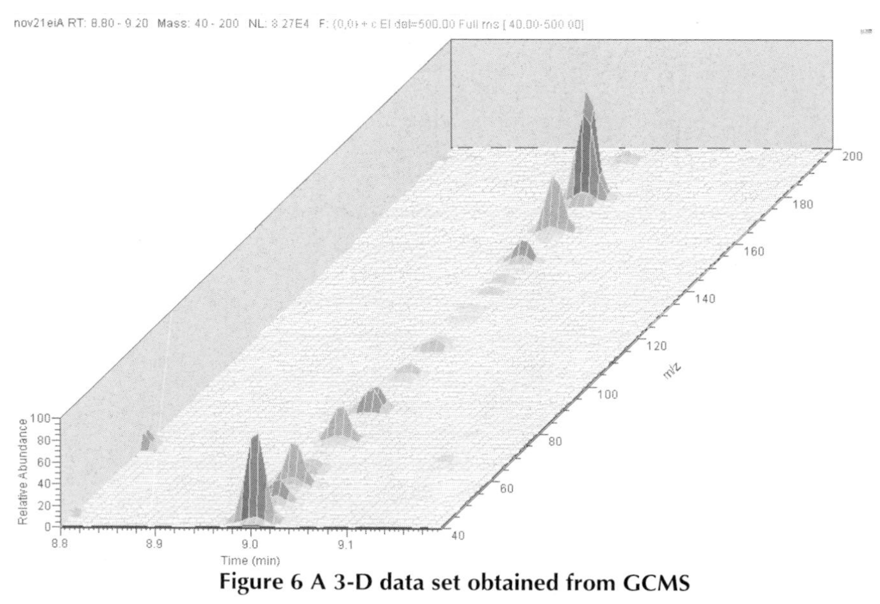

Figure 6 A 3-D data set obtained from GCMS

If a drug is detected, another sample of the urine is taken for confirmatory analysis.

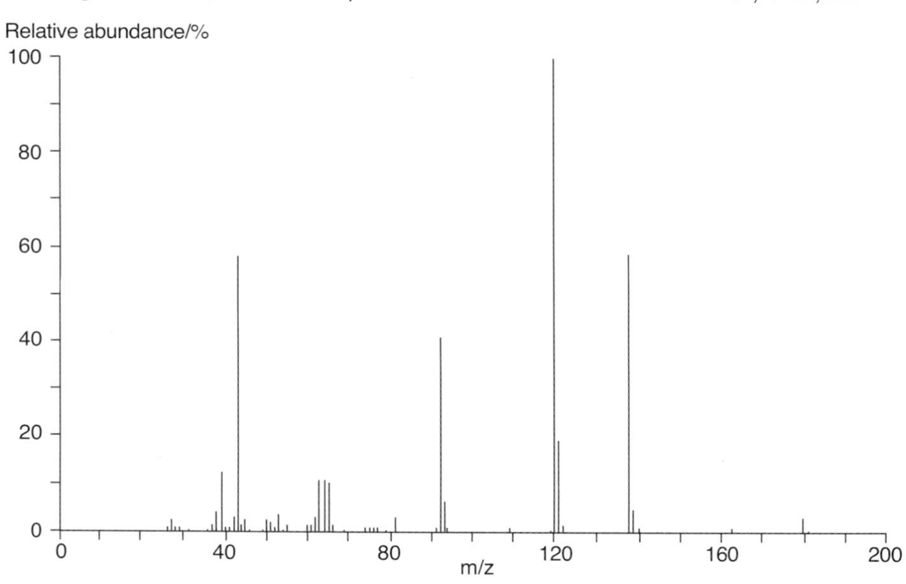

Figure 7 The mass spectrum of aspirin

Figure 8 The displayed formula of aspirin

RS•C

Q5. Figure 7 shows the mass spectrum of aspirin and Figure 8 its displayed formula.

a) i) Use the displayed formula to help you suggest what fragments of the aspirin molecule are responsible for the peaks at mass 43 and mass 121.
ii) What causes the peak at mass 180?

b) Suggest why a racehorse might be given aspirin.

Confirmatory analysis

Many of the screening processes described above are carried out automatically by the instruments themselves under computer control. This means that no single analyst will have supervised the whole screening process for any single sample. If a regulatory authority, such as The Jockey Club, wishes to take disciplinary proceedings as a result of drug testing, it is vital that the analysing laboratory is able to state without fear of contradiction that:

1. There was no way in which samples from different horses had been mixed, confused or contaminated.
2. The materials claimed to be present were actually present.
3. The instrumentation used was not giving false positives (indications that substances were present when they actually were not).

To ensure that this is the case, a further analysis is carried out. This time all the operations are carried out or supervised by a single analyst. This analyst will then be able to give evidence in a court of law, if required.

The analyst prepares a reference sample that contains known drugs to check that the analysis method detects them and so that the mass spectrum of this sample can be compared with that from the horse. He or she then performs the analysis on four samples:

1. a system blank consisting of water or a buffer solution to show that there is no contamination from one liquid to another and that the instrumentation is not giving false readings;
2. a biological blank – urine or blood plasma as appropriate – which is known to be drug free;
3. the sample believed to contain the drug identified during screening; and
4. the reference sample.

In general, the regulatory bodies are interested only in whether a drug substance is or is not present rather than its concentration. Complications may arise if the substance found could reasonably be expected to be present in minute concentrations because it is a naturally occurring compound, *eg* theobromine or salicylic acid (2-hydroxybenzoic acid). In such cases the regulatory authority prescribes a threshold limit.

Action taken when a test is positive

Once the analyst is satisfied that the positive result from screening is correct, a case file is prepared. This contains the screening tests – immunoassay results, the gas chromatogram and mass spectra – and the confirmatory data. This is checked by another analyst before the report is sent to the regulatory authority, which then decides on the action to be taken.

Few cases of doping are reported in this country and this shows the care with which owners and trainers follow the rules of their regulatory authorities. Table 1 shows some recent figures.

RS•C

	1997	1998	1999
samples analysed	7,596	7,622	7,764
positive test results	18	5	21
% of tests being positive	0.24%	0.07%	0.27%

Table 1 Samples tested and positive results found for horses tested in 1997–1999. The number of horses tested represents about 10% of those racing.

(Data from **http://www.thejockeyclub.co.uk/jockeyclub/html/racing/antidope.htm**, accessed October 2001.)

It should be added at this stage that not all cases where drugs are found are deliberate attempts to alter the outcome of a race. Horse owners and trainers occasionally give their horses medications, feedstuffs and supplements that contain ingredients that they did not know were banned, or which are there as contaminants whose presence could not be known until the horses were dope tested.

An example of this was the disqualification of the British showjumper David Broome and his horse, Lannegan, in the 1990 Irish Nations Cup. As was reported in the media at the time, the British team won the trophy and were going to retain it permanently because this would have been the third consecutive win. When Lannegan was tested, a substance called isoxsuprine was detected. Isoxsuprine is routinely used to treat navicular disease, a condition in which the flow of blood to the foot is impaired. It dilates (opens up) blood vessels, improving the circulation of blood and can enable a lame horse to run. This is why it is banned. Investigations followed, which unearthed the fact that a rehydrating drink that had being given to the horse in the days before the race also contained isoxsuprine. The company that manufactured the drink, which should contain only salts and sugar, sent another sample of the drink from the same batch for analysis. This also proved positive. Further inquiries revealed that a shovel used in the department making the drink had also been used in a part of the factory manufacturing isoxsuprine, and that enough had been transferred to the drink to cause the horse to fail a drug test. The manufacturing company that made the drink paid David Broome the prize money that he had lost (see *The Human Element*, John Emsley, BBC, 1992).

Figure 9 David Broome and Lannegan competing at Hickstead, 1989
(Reproduced by courtesy of Bob Langrishe)

Detecting drugs in horse urine by immunoassay

Every horse (and every human) has an immune system whose job is to recognise 'invaders' that are not part of that individual. This is mainly to protect the individual against micro-organisms such as bacteria and viruses. However, the immune system will also recognise 'foreign' molecules, including drugs. The 'foreign' molecules are called antigens in this context. On recognising a 'foreign' molecule, the immune system produces a molecule called an antibody that bonds specifically and irreversibly to this particular antigen and this antigen only. You can think of the two molecules fitting together like a key in the right lock. An antibody will bond only to its particular antigen. This process is called the immune response.

Immunoassay is a method of detecting and measuring the concentrations of particular drugs in urine by using this highly specific immune response – the bonding between an antigen (the drug) and its antibody. The Horseracing Forensic Laboratory (HFL) has a range of antibodies that are specific for a particular drug or drug metabolite molecule (one of the molecules that drugs are converted into by the liver). If the antibody and its antigen drug molecule meet, they will bind together very strongly in a 1:1 ratio.

The advantage of the method is that there is no need to separate the drug that is to be measured from the cocktail of other chemicals in the horse urine sample; the antibodies effectively 'fish out' their own antigen – the drug molecule – from even the most complex mixture.

The ELISA screening technique

This method is widely used in biomedical science, including one of the tests for Human Immunodeficiency Virus (HIV). Its full name is Enzyme Linked Immuno Sorbent Assay (ELISA), and HFL uses it to screen urine samples for the presence of drugs, alongside the Gas Chromatography-Mass Spectrometry (GCMS) method (see *Identifying drugs in horse urine by Gas Chromatography – Mass Spectrometry* (GCMS)). The ELISA method can test urine without a lengthy preparation, thus speeding up the analysis process considerably. The test is also relatively cheap and quick to use.

To test for a particular drug, a polythene microtitre plate (see Figure 1) is prepared. Its small test wells are lined with the protein Bovine Serum Albumin (BSA) that has the drug molecule covalently bonded to it. Each microtitre plate has 96 wells, and HFL have a range of plates, each with a particular drug (or drug metabolite) in place, Figure 2.

RS•C

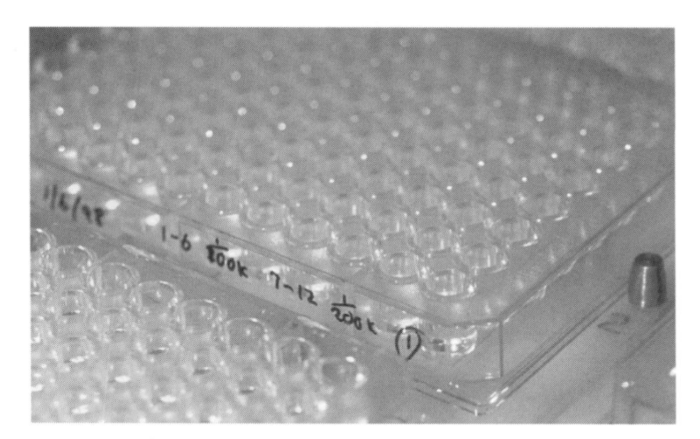

Figure 1 An 8 x 12 microtitre plate

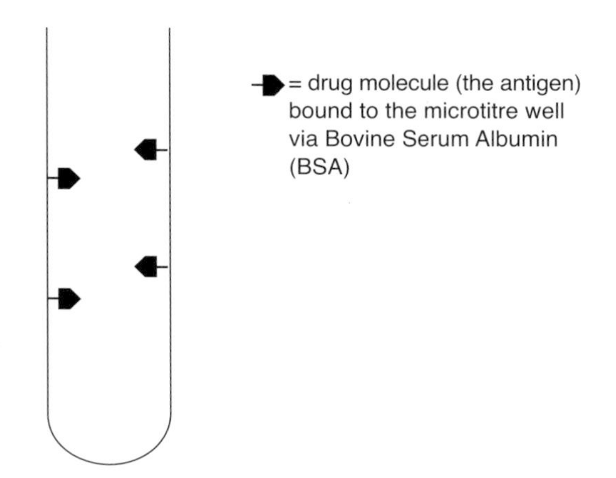

Figure 2 A well with bound drug

Each plate can accommodate 90 different urine samples as well as three controls at the beginning and end that are used as a check on the accuracy of the analysis. Each plate is processed on a computer-controlled machine (Figure 3) that can be programmed to add a variety of reagents in any desired order. Each urine sample is identified by a bar code, and its position on the test plate is recorded automatically by the computer.

Figure 3 Processing microtitre plates

RS•C

The sequence of events in the analysis is described below. Each of Figures 4 to 9 has two parts, (a) which represents a negative urine sample (one not containing the drug in question) and (b) which represents a positive urine sample (one containing the drug in question).

Step 1
The sample of untreated urine is added, followed by antibody, Figure 4.

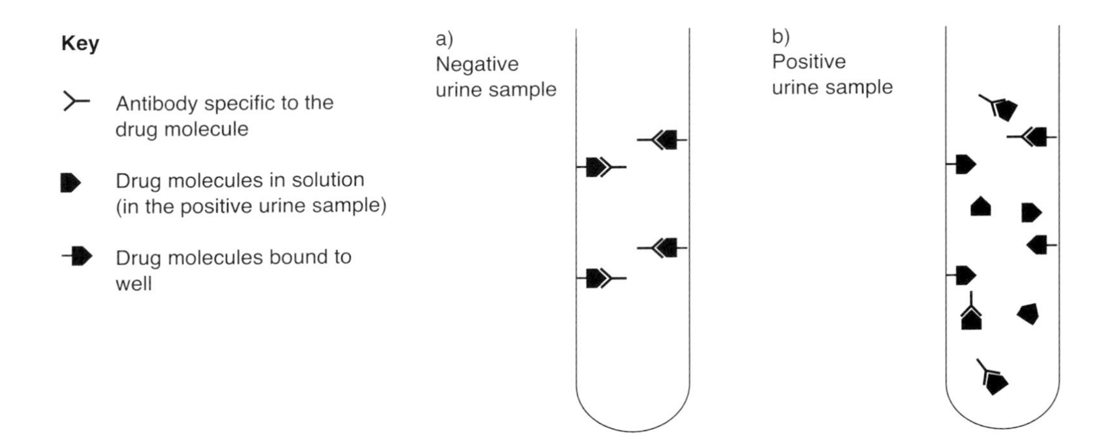

Figure 4 Urine and antibody are added to each well

The microtitre plate is then gently shaken at 37 °C, and the antibody binds irreversibly to molecules of the drug. This includes both drug molecules bound to BSA on the wall of the well and drug molecules in solution. The amount of antibody added is just enough to bind with the drug molecules bound to BSA on the wall of the well. So, in a negative urine sample, all the drug molecules attached to the wall will bind with antibody. In the positive urine sample, some antibody will bind to drug molecules attached to the wall and some will bind to free drug molecules in the sample. This type of immunoassay is known as a 'competitive' method because the bound and free drug molecules compete for the antibody.

Q1. a) Why is it important that the sample is introduced to the test well before the antibody is?

b) There will be hundreds of compounds in any urine sample; how does this method ensure that only the target drug is detected?

Step 2
The wells are then washed.

This removes any drug-antibody complex that is not attached to the well, Figure 5.

RS•C

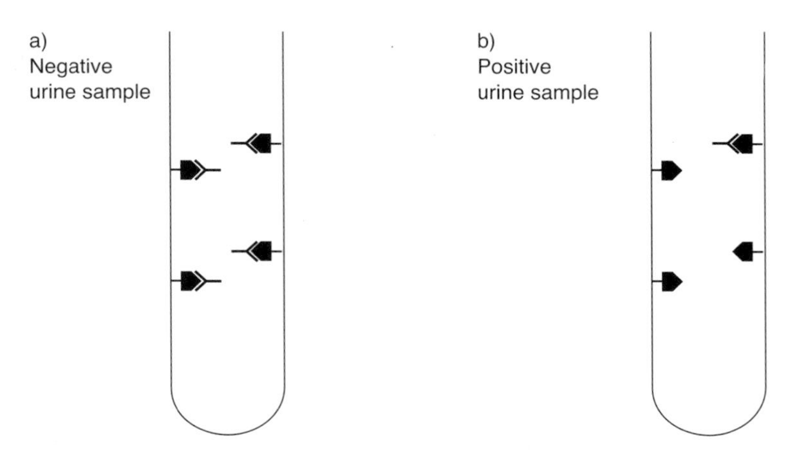

Figure 5 The well is washed

Step 3

Next, an enzyme, called alkaline phosphatase, is added that bonds to the antibody via a protein molecule. The negative sample will have much more enzyme bound to the antibody than will the positive sample, Figure 6.

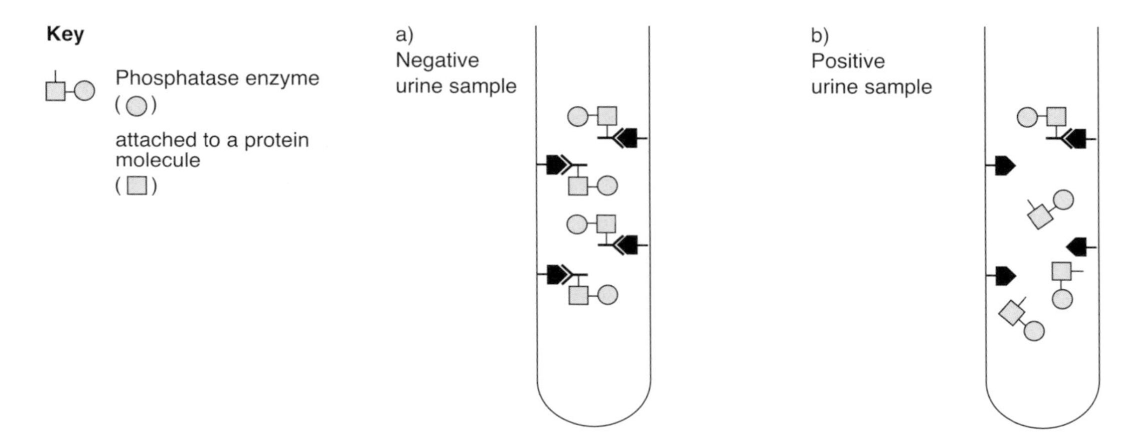

Figure 6 The phosphatase enzyme is added

Step 4

The wells are washed again. This removes the unbound enzyme, leaving only the enzyme bound to the wall via the drug and antibody molecules. The negative sample has much more bound enzyme than the positive sample, Figure 7.

RS•C

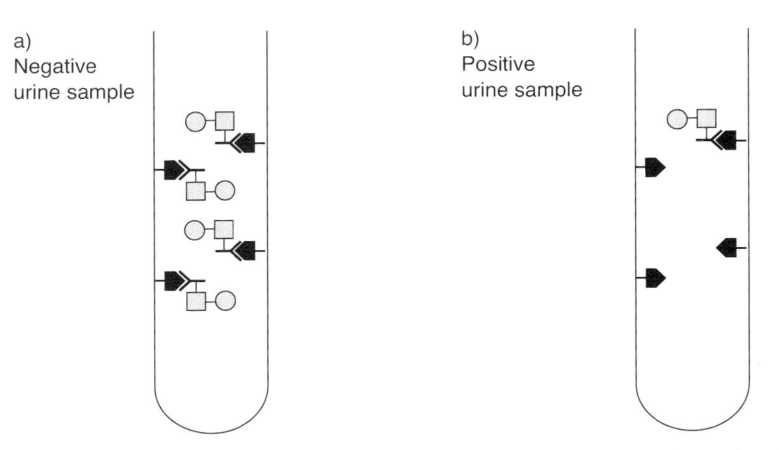

Figure 7 The negative urine sample has much more enzyme bound to it than the positive one

Step 5

The enzyme's substrate (4-nitrophenylphosphate) is added in a buffer solution of pH 9.6 based on the molecule diethanolamine, see Figure 8. The mixture is allowed to react for 30 minutes at 37 °C.

Key

○ Enzyme substrate 4-nitrophenylphosphate

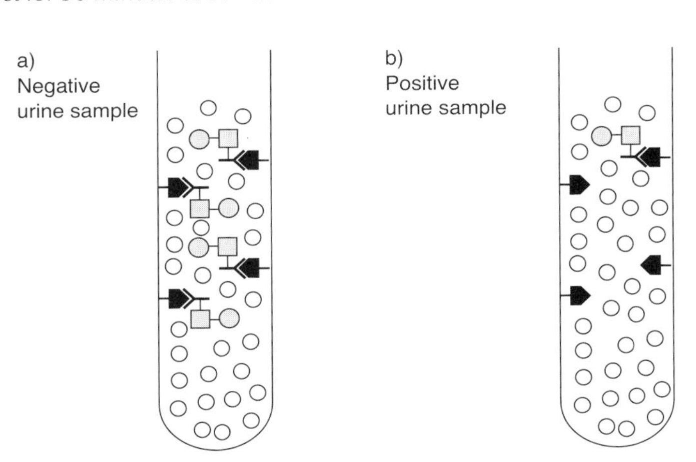

Figure 8 The enzyme substrate, 4-nitrophenylphosphate, is added

The reaction brought about by the enzyme is:

NO_2 — C_6H_4 — O — PO_3^{2-} $\xrightarrow{\text{alkaline phosphatase}}$ NO_2 — C_6H_4 — OH + HPO_4^{2-}

4-nitrophenylphosphate (colourless)

4-nitrophenol (yellow)

RS•C

Q2. Diethanolamine has the structure shown below.

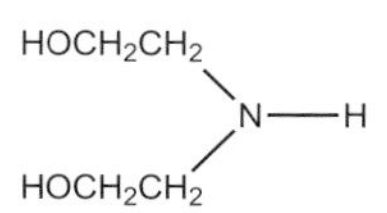

a) **Diethanolamine is a base in aqueous solution. Explain how the structure of this molecule allows it to behave as a base and give an equation to show its basic behaviour.**

b) **What is a buffer?**

c) **Diethanolamine alone is unable to act as a buffer. Suggest a second compound that could be mixed with diethanolamine to produce a buffer.**

d) **Why is a buffer essential for this step of the immunoassay procedure?**

e) **What type of chemical reaction is the reaction of alkaline phosphatase on 4-nitrophenylphosphate?**

Step 6

The product of the enzyme-catalysed reaction, 4-nitrophenol, is yellow. The amount of yellow product will be high in the wells containing negative samples and there will be little or no yellow colour in positive sample wells. This gives the test result – a negative sample (one containing no drug) will be yellow and a positive sample (one containing the drug) will be almost colourless, Figure 9. The paler the yellow colour, the greater the concentration of the drug in the original urine sample.

Key

○ 4-nitrophenol

a) Negative urine sample

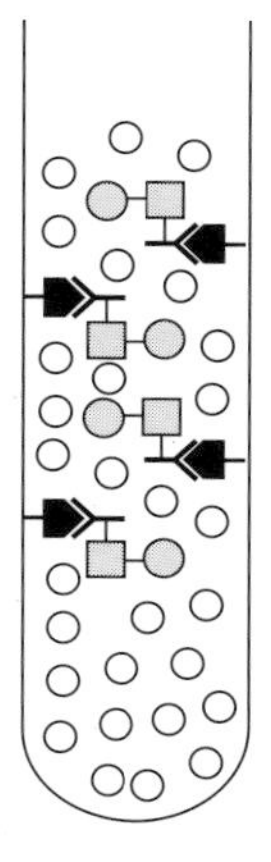

b) Positive urine sample

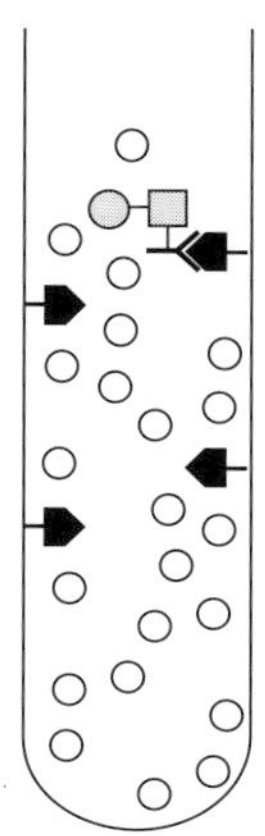

Figure 9 4-nitrophenol is yellow in colour, so the negative urine sample appears darker yellow than the positive sample due to a higher 4-nitrophenol concentration

Step 7

The amount of light of wavelength 405 nm absorbed by the solution is measured and recorded. This is called the absorbance of the solution and effectively measures the depth of the yellow colour. Positive results usually look colourless and can be detected by eye. However the absorbance at 405 nm of all the wells is measured automatically by a spectrophotometer (a sophisticated colorimeter).

RS•C

By comparing the absorbances of the blank and control wells with that of the positive sample, the computer can estimate the amount of drug present in each sample well. This will be confirmed by the results from Gas Chromatography-Mass Spectrometry (GCMS) for the same drug. (See *Detecting drugs in horse urine by Gas Chromatography Mass Spectrometry* (GCMS).)

Q3. Light with a wavelength of 405 nm is in the blue region of the visible spectrum. Why is this wavelength chosen to measure the concentration of 4-nitrophenol, which is yellow in colour?

Q4. Make a simple sketch that shows the essential components of a colorimeter used to measure the amount of light absorbed by the yellow solution. Briefly describe what each component does.

Drug metabolism

Finding out if an animal (including a human) has taken a drug is not quite as simple as just analysing for the drug itself in body fluids such as blood or urine. When an animal takes in a 'foreign' chemical such as a drug, the body's chemical machinery begins to get rid of it, as quickly as possible. (This is why the effect of an aspirin, for example, wears off after a few hours.) The original drug molecules are converted into a number of different molecules that are more easily excreted into the urine. These molecules are called metabolites and the whole process is called metabolism. So, depending on when the drug was taken, the analyst may have to look for the drug itself and/or its metabolites. The liver is the main organ where chemical changes to the drug take place, and the kidneys are mainly responsible for the excretion of the foreign chemical into the urine.

The metabolism of the drug will begin as soon as it is administered and in many cases the process follows first order kinetics, *ie* the rate of metabolism is proportional to the concentration of the drug.

As a result, the amount of drug in the animal's bloodstream usually decays away as shown in Figure 1, showing a constant half-life. The half-lives of drugs vary enormously, some take months to be eliminated from the body and some just a few hours.

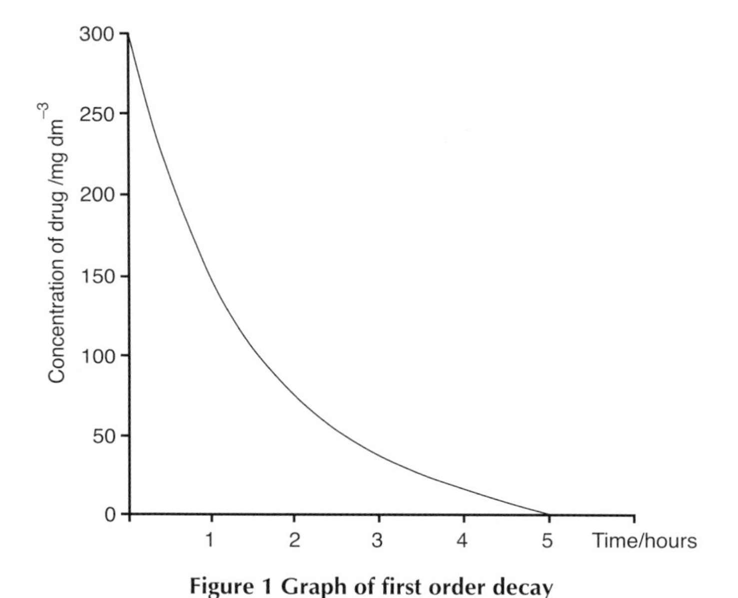

Figure 1 Graph of first order decay

Q1. **a)** Use the graph in Figure 1 to work out the first three half-lives of the decay and show that they are the same.

b) If a drug has a half-life in a horse of 12 hours, by how much has the drug's concentration dropped in 2 days?

Animals may be given potent drugs for legitimate reasons. Horseracing is a global activity with prestigious and valuable races held worldwide and racehorses are often carried by air. So acepromazine (Figure 2), a sedative, is frequently given to racehorses to calm them down before they are transported by air.

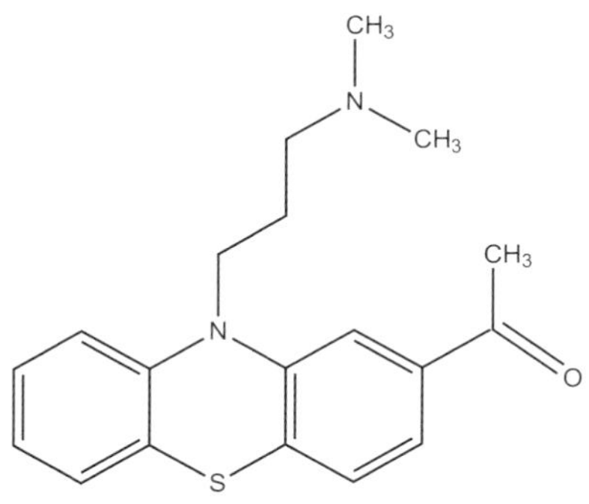

Figure 2 Acepromazine

Trainers of racehorses must be careful to allow sufficient time for a drug such as this to be metabolised out of the animal's system before a race, as it would show up on an analysis of urine or blood taken afterwards. A knowledge of how long drugs remain in the system of an animal is clearly very useful for vets and anyone involved in animal welfare, particularly if the animals are involved in competition.

Enzymes are involved in the metabolism of drugs. One enzyme normally brings about one specific reaction only, but, unusually for enzymes, a group of enzymes collectively known as cytochrome oxidase P450 is able to tackle a range of substrate molecules.

The process of drug metabolism usually occurs in two steps:

1. A step involving oxidation, hydrolysis, reduction, sulfonation (substitution of a hydrogen by an $-SO_3H$ group), or de-alkylation (removal of an alkyl group such as $-CH_3$ or $-C_2H_5$). These are called phase I reactions.
2. A step involving reaction with glucuronic acid (a sugar-like molecule) or sulfate. This step tends to remove or mask some of the functional groups in the molecule. These are known as phase II reactions.

Q2. Compare the structure of glucuronic acid (Figure 3) with that of glucose (Figure 4). Point out any similarities and any differences.

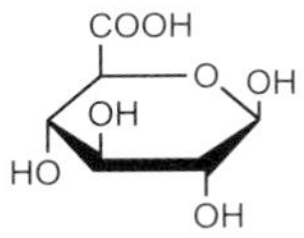

Figure 3 Glucuronic acid

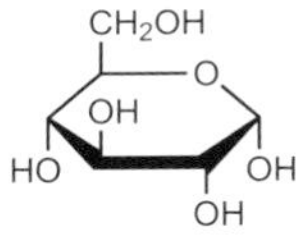

Figure 4 Glucose

Phase I reactions

Examples of phase I reactions include:

- oxidation of phenyl groups via an epoxide (an epoxide is a molecule containing a ring of two carbon atoms and one oxygen atom) to phenols (Figure 5)

RS•C

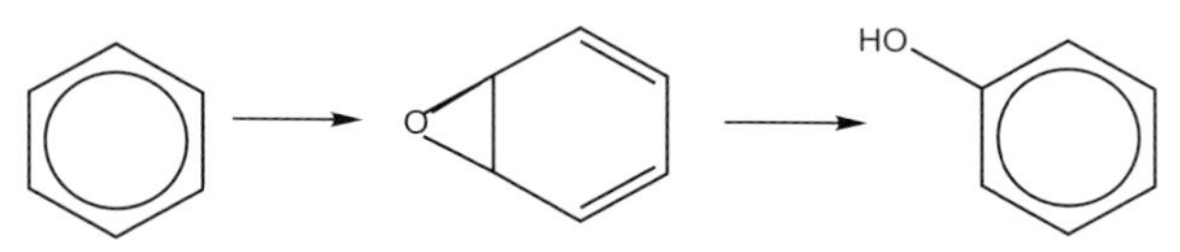

Figure 5

- reduction of carbonyl groups to alcohols (Figure 6)

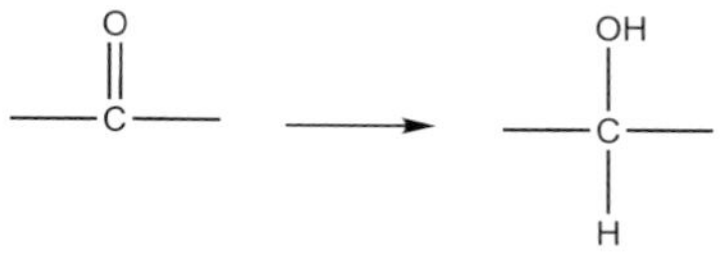

Figure 6

- removal of the R groups from N–R and O–R (where R represents an alkyl group such as CH_3 or C_2H_5) (Figure 7)

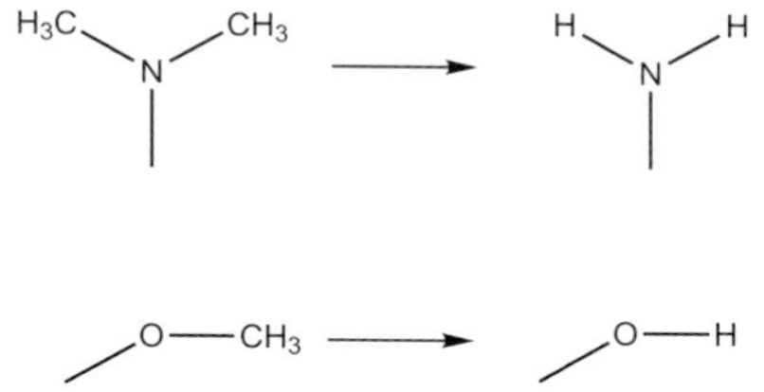

Figure 7

Q3. a) Draw an epoxide ring.

b) What are the approximate C–O–C and C–C–O angles in an epoxide?

c) What are the approximate angles of these bonds normally?

d) What does this suggest about the stability of the epoxide group?

Q4. How will each of the phase I changes in the examples above affect the water solubility of the molecules? Explain your answers.

Q5. How will the phase I changes in the examples above help the removal of the metabolised drug from the body?

Q6. The phase I metabolism of acepromazine (ACP) can occur via two routes as shown in Figure 8.

What types of reaction are a, b, c and d?

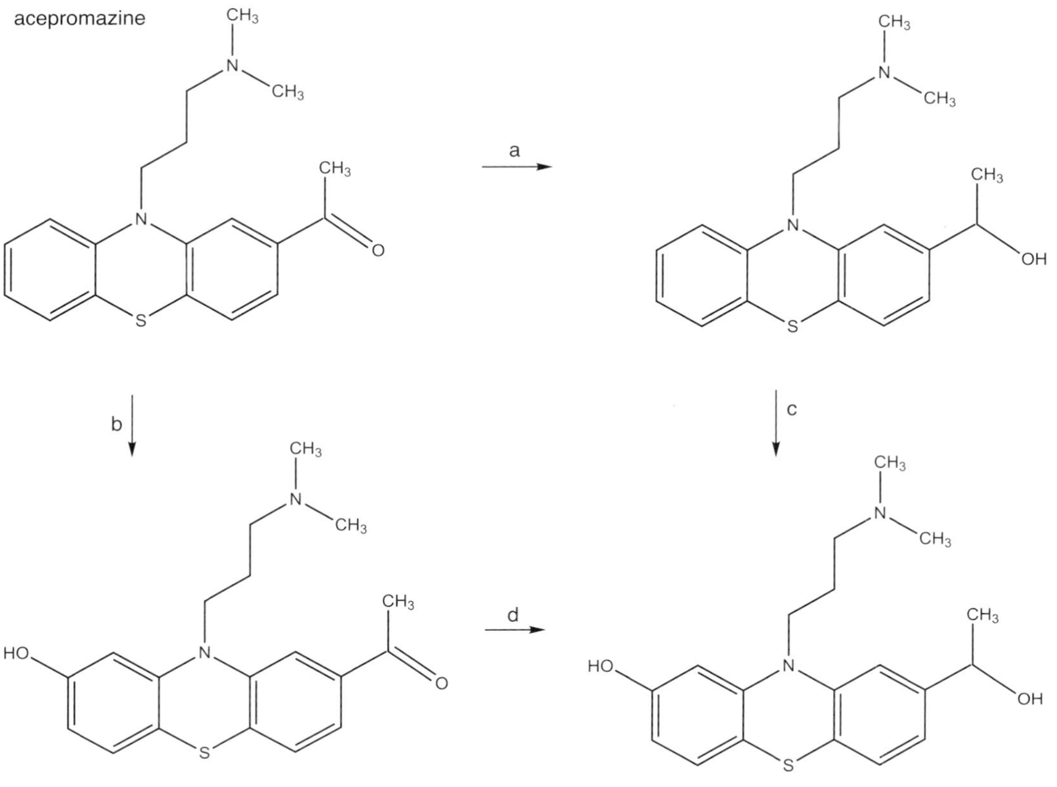

RS•C

Figure 8 Metabolism of acepromazine

> **Q7. Which other functional groups of the acepromazine molecule might be susceptible to metabolic change by phase I reactions?**

Phase II Reactions

The product of the reaction of glucuronic acid with acepromazine metabolite is shown in Figure 9. This product is termed a glucuronide.

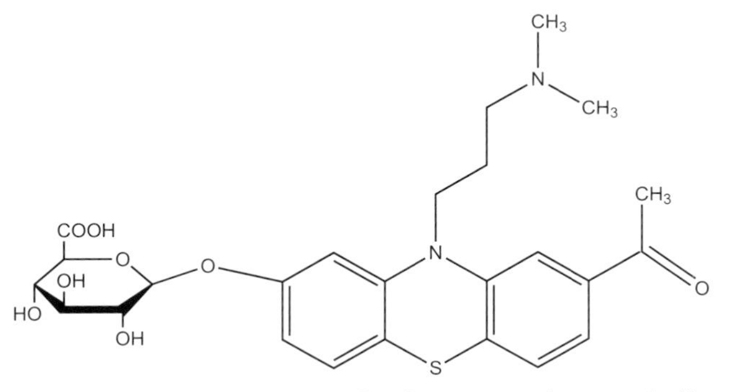

Figure 9 The glucuronide of acepromazine metabolite

> **Q8. a) How will the formation of the glucuronide affect the water-solubility of the metabolite? Explain your answer.**
>
> **b) How will this help the excretion of the drug metabolite from the body?**

Drug metabolites in forensic analysis

Analysis of fluids, such as blood or urine, for evidence of drugs is always complicated by the many natural chemical components in them. Substances that are normally present in these fluids are called endogenous materials, while those that do not occur naturally are described as exogenous. The analysis for exogenous compounds in blood or urine is clearly complicated by the presence of large quantities of endogenous ones.

Q9. Suggest three endogenous chemical components of blood or urine.

The fact that a drug molecule is changed radically by the body as it is metabolised, means that a forensic analyst looking for an exogenous substance in the animal may find several drug metabolite molecules, but not the drug itself. The interpretation of the findings can therefore be complex. The analyst is often faced with the problem of having found metabolites that could have derived originally from a number of different drugs.

For example, the mild over-the-counter painkiller codeine is metabolised in the body to morphine (see Figure 10)

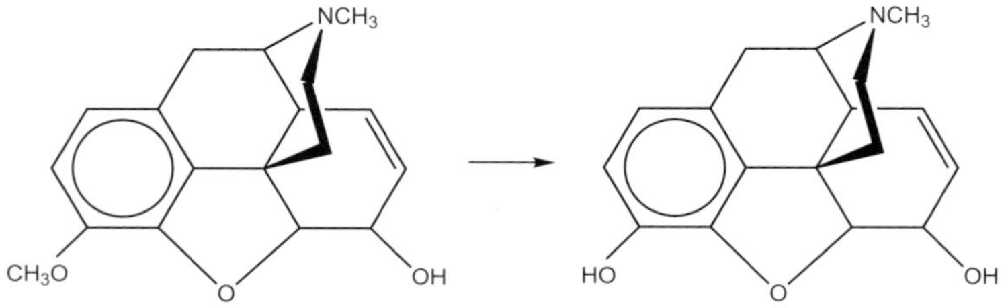

NCH_3 NCH_3

CH_3O O OH → HO O OH

codeine morphine

Figure 10 Codeine is metabolised into morphine

Q10. What type of metabolic reaction has occurred here?

However, the much more powerful and addictive drug heroin is also metabolised into morphine, see Figure 11.

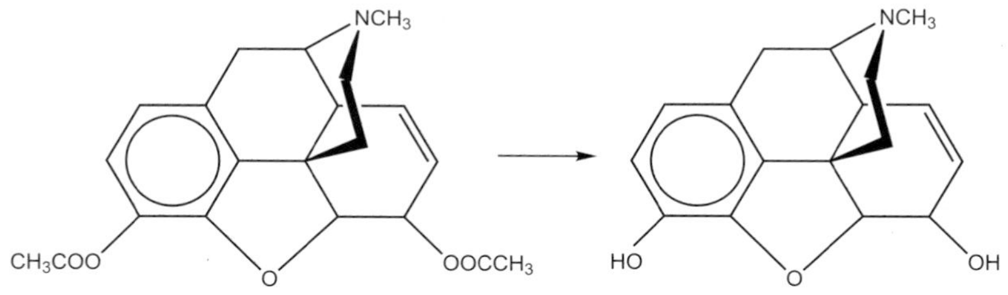

NCH_3 NCH_3

CH_3COO O $OOCCH_3$ → HO O OH

heroin morphine

Figure 11 Heroin is also metabolised into morphine

So if morphine is found in blood or urine, at first sight it seems difficult to decide whether this is evidence that morphine itself, codeine or heroin has been taken. However, the pathway from heroin to morphine proceeds via the short-lived monoethanoyl morphine (Figure 12), and the presence of the latter in blood or urine would be good evidence of heroin having been given as a drug originally, rather than codeine.

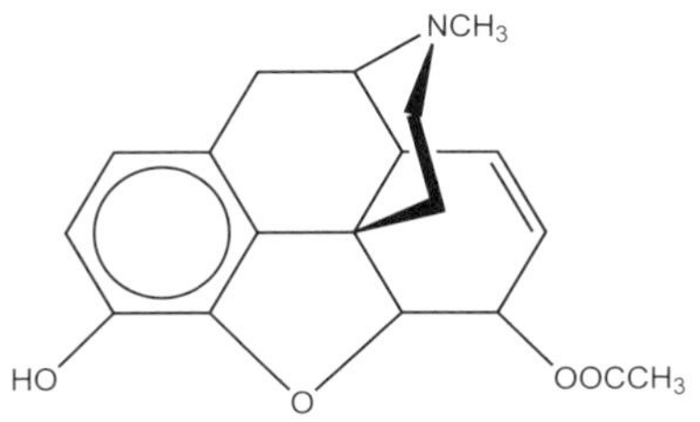

NCH_3

HO O $OOCCH_3$

Figure 12 Monoethanoyl morphine

The fact that phase II type reactions often produce metabolites carrying extra residues such as glucuronic acid poses a problem for the analyst. The first step in an analytical procedure is usually to remove the glucuronic acid using enzymes before carrying out the analysis.

Identifying drugs in horse urine by GCMS

The urine of a horse contains (as well as water) a cocktail of many hundreds of chemical compounds including inorganic salts as well as organic compounds, some of them quite complex. If a horse has been given a drug, some of this may be excreted unchanged in the urine. It is also likely that much of the drug will have been metabolised, that is chemically changed into related compounds called metabolites.

To determine whether a particular drug (or its metabolites) is present in urine requires this mixture to be separated and the drug molecule and/or its metabolites to be identified.

When the Horseracing Forensic Laboratory (HFL) receives a sample, it may already have been split into two parts – one is kept frozen for later analysis to confirm the results if required, possibly in another laboratory. A portion of the sample to be analysed is also split into two – one part is investigated by a technique called immunoassay that uses antibodies to detect specific groups of drugs (see *Identifying drugs in horse urine by immunoassay*). The second part of the sample is investigated by Gas Chromatography-Mass Spectrometry (GCMS). This is really two techniques used together in which gas chromatography (GC) is used to separate the components of the mixture and the mass spectrum (MS) of each separated component is generated to help identify it. The two instruments are coupled together so that as each component is separated, it is fed straight into the mass spectrometer. This generates a massive amount of data – potentially hundreds of components of the mixture each with a mass spectrum with many lines.

Sample preparation

Before GCMS is used the urine is treated in two ways – firstly it is placed in a buffer solution and treated with enzymes. The enzymes reverse some of the chemical changes that have occurred to the drug in the horse's liver. Then some of the organic substances (including drugs and their breakdown products, called metabolites) are separated from the inorganic salts in the urine by a process called solid phase extraction. This involves pouring the treated urine sample through a tube containing a powder called silica. This absorbs the drug molecules but not the inorganic salts. The solid is then washed with a number of liquids to remove as much interfering material as possible but leave behind any drugs. Two different solvents are then poured through the tube to wash out the drug molecules. One of these washes out acidic and neutral drugs, the other washes out any basic drugs. Each category of drug is then analysed by GCMS individually.

Q1. What is a buffer solution? Suggest why it is used in this process.

Acidic and neutral drugs include:

- non-steroidal anti-inflammatory drugs, including mild pain-killers such as aspirin;
- diuretics, given to lower blood pressure; and
- methylxanthines (mild stimulants such as caffeine).

Basic drugs include:

- sedatives;
- local anaesthetics;
- β-blockers, to slow the heart rate;
- stimulants;
- narcotics, such as morphine; and
- β-agonists, to improve breathing.

Q2. a) Figure 1 shows the structure of aspirin, a non-steroidal anti-inflammatory drug. Suggest which part of the molecule makes it acidic.

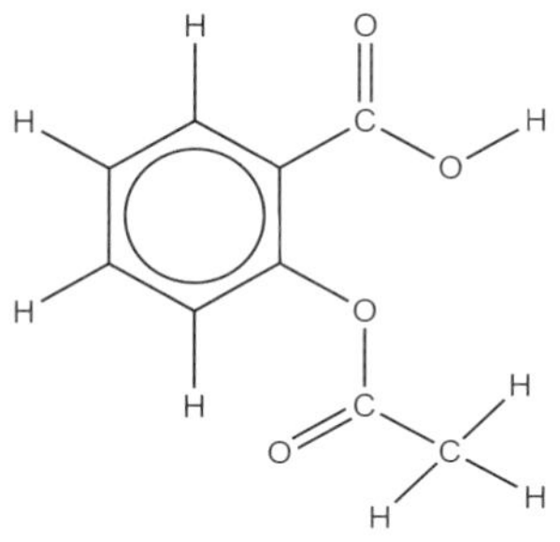

Figure 1

b) Figure 2 shows the structure of acepromazine, a sedative drug often given to horses to calm them down when travelling. Suggest which part of the molecule makes it basic.

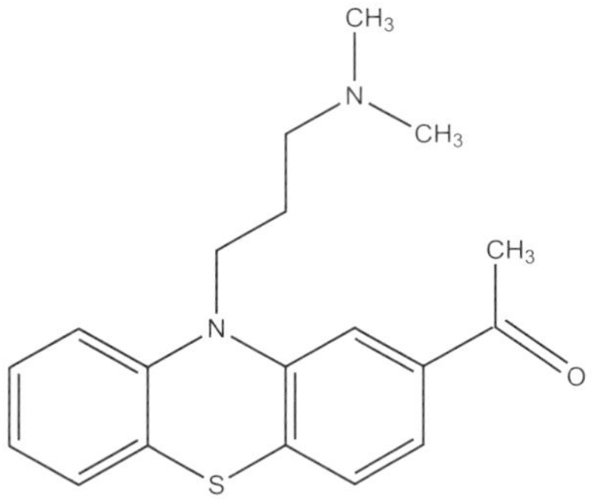

Figure 2

Gas Chromatography – Mass Spectrometry

Gas chromatography (GC)

This works on the same principle as paper chromatography. In paper chromatography a liquid flows over paper and carries the components of a mixture along with it, each component moving at a different rate. The paper is called the stationary phase and the liquid the mobile phase.

The heart of the GC process is a 25 m-long thin tube made of silica coated with other chemicals. This tube is so thin (internal diameter about 0.25 mm) and flexible that it can easily be tied in knots. The tube, called a 'column', is wound into a spiral and sits in an oven whose temperature can be controlled, see Figure 3. A stream of helium gas flows through the column. A sample of 1 μdm^3 of one of the mixtures resulting from solid phase extraction is injected into the column. The components of the mixture are carried through the column by the stream of helium, taking between 10 and 20 minutes to travel the 25 m. This separates the mixture because some components are carried through with the helium more quickly than others, which tend to 'stick' to the coating of the column. The 'stickier' the component, the longer it takes to pass through the column. The time taken for any particular component to pass through the column is called the retention time. As each component of the mixture leaves the column its amount is measured by a detector.

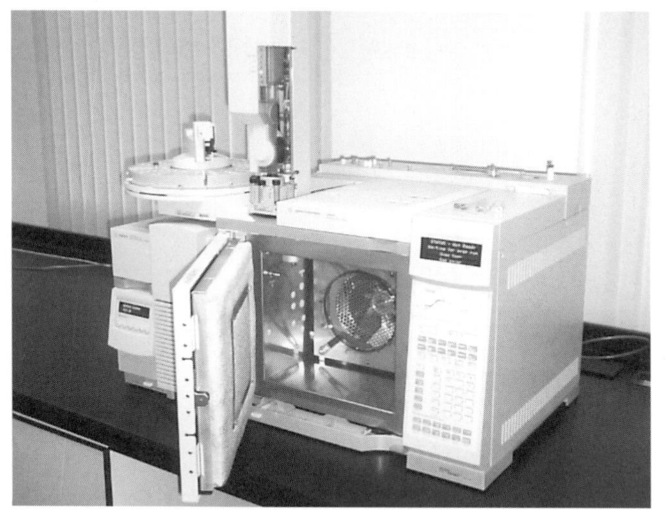

Figure 3a Agilent 5973N GCMSD instrument, showing column inside oven

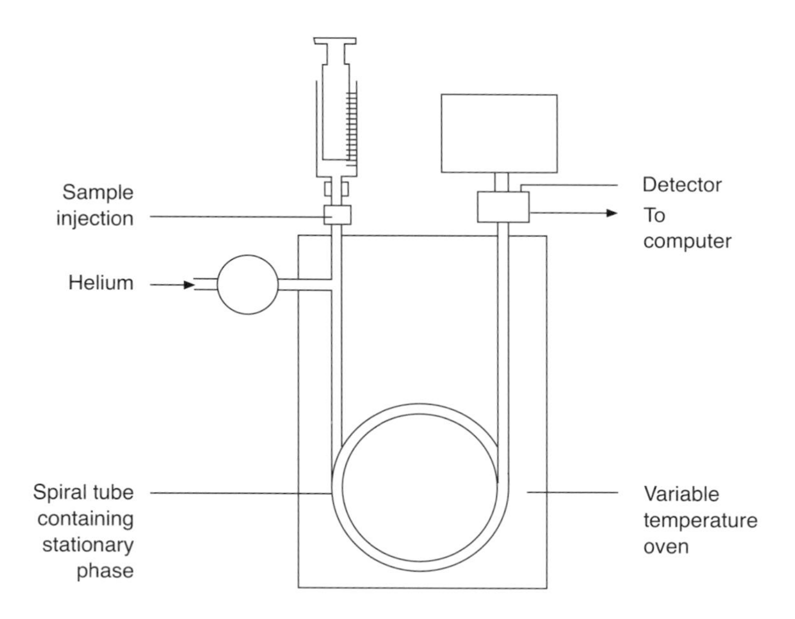

Figure 3b schematic of GC instrument

RS•C

> **Q3. Suggest what sorts of forces are responsible for the molecules 'sticking' to the column?**
>
> **Q4. GC can be compared to paper chromatography, in which water soaks along filter paper and separates a mixture of coloured dyes. In GC, what corresponds to the paper (stationary phase) and what corresponds to the water (mobile phase)? What corresponds to the distance each component moves along the paper?**

The output of the detector is fed into a computer that can plot a graph of detector signal against retention time. This is called a chromatogram. A typical one is shown in Figure 4.

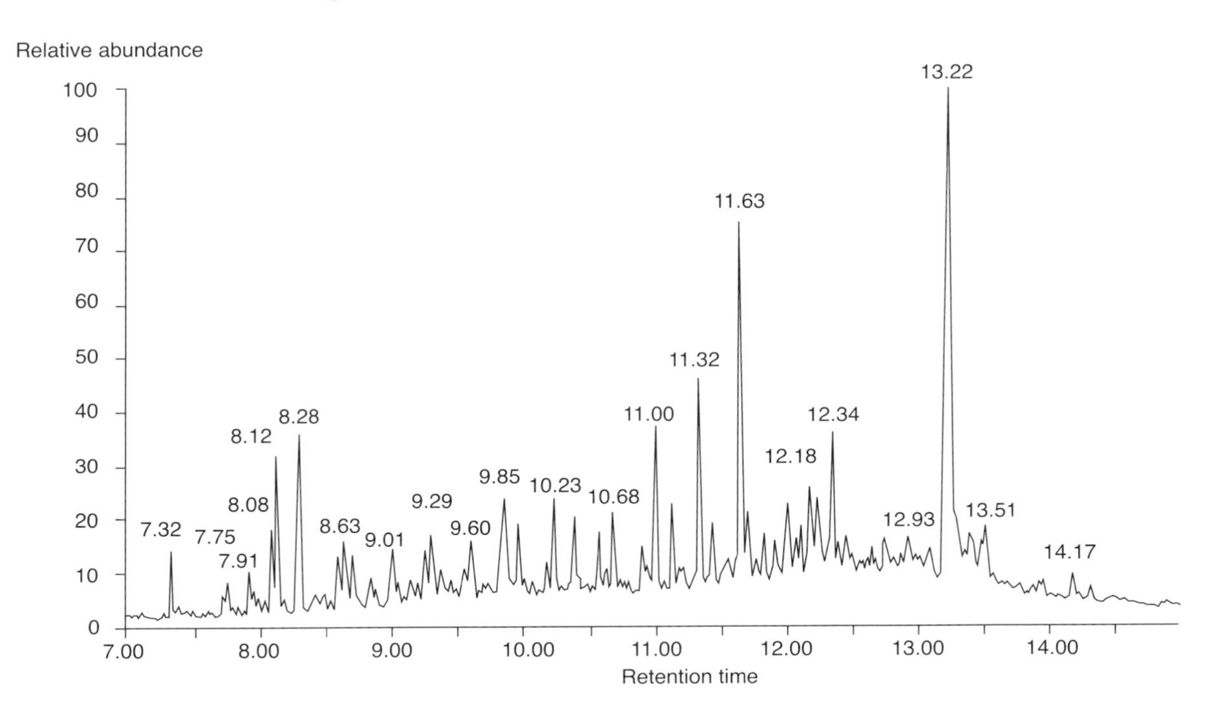

Figure 4 A typical gas chromatogram

Each peak in the chromatogram corresponds to a different component of the mixture. The taller the peak (strictly speaking its area), the more of that component there is in the mixture. In Figure 4, the numbers printed by some of the peaks are the retention times of those peaks in minutes. The computer used to store, manipulate and print the data has been used to mark the retention times of the main peaks of the chromatogram to make it easier to interpret.

RS•C

Any particular compound will always take the same amount of time to travel through the column (providing the instrument conditions are the same). This means that the retention time of a peak in the chromatogram gives an indication of what substance that peak represents.

Q5. Suggest what conditions, if changed, might affect the retention time.

So, for example, in Figure 4, the analyst might spot the peak with retention time 9.14 minutes and suspect that it represented the compound pethidine because he or she knows that the retention time for pethidine under these conditions is 9.14 minutes. Pethidine is a painkiller and might be given to a racehorse to enable it to run while it had an injury. To prevent this happening, horses in the UK are not allowed to race while they have this substance in their systems.

Because conditions affect the retention time, it is not sufficient to measure the retention time to be certain of the presence of a particular drug. Analysts have developed a trick to help them – they measure the so-called relative retention time of each component. Each sample is 'spiked' with a small amount of a substance called an internal marker (IM). The peak in the chromatogram corresponding to the IM is identified and the retention times of other components can be measured relative to this. A compound with a relative retention time of 1.1, for example, would take 1.1 times longer than the IM to travel through the column. This takes account of any factors that might slow down or speed up the passage of all the components through the column.

However, even the presence of a peak with a relative retention time the same as that of a known drug (or metabolite) would not be enough on its own to establish with certainty that that drug or metabolite was present in the sample. It is possible that, coincidentally, a drug molecule might have the same retention time as another entirely unrelated compound. To establish the presence of a drug without doubt, the analyst must look at the mass spectrum.

Mass spectrometry

In simple GC, a variety of detectors are used. These simply tell the analyst when a component has reached the detector and how much of it there is. In GCMS, the components leaving the column are fed directly into a mass spectrometer. This can do more than simply detect the arrival of a component and measure its amount, it can help to identify it.

In the mass spectrometer, the molecules of the sample are first bombarded with a beam of electrons. This has two effects. Firstly an electron is knocked out of a sample molecule so that the molecule becomes a positive ion. The loss of the electron makes the resulting ion unstable. Secondly, because of the energy of the collision, some of the bonds in the ions break so that the sample breaks into fragments. The inside of the spectrometer is kept under a high vacuum and these fragment ions fly through the mass spectrometer under the influence of electric and magnetic fields and are separated by their masses. A mass spectrum of each component is produced like that in Figure 5. This is essentially a graph of amount against mass of fragment.

RS•C

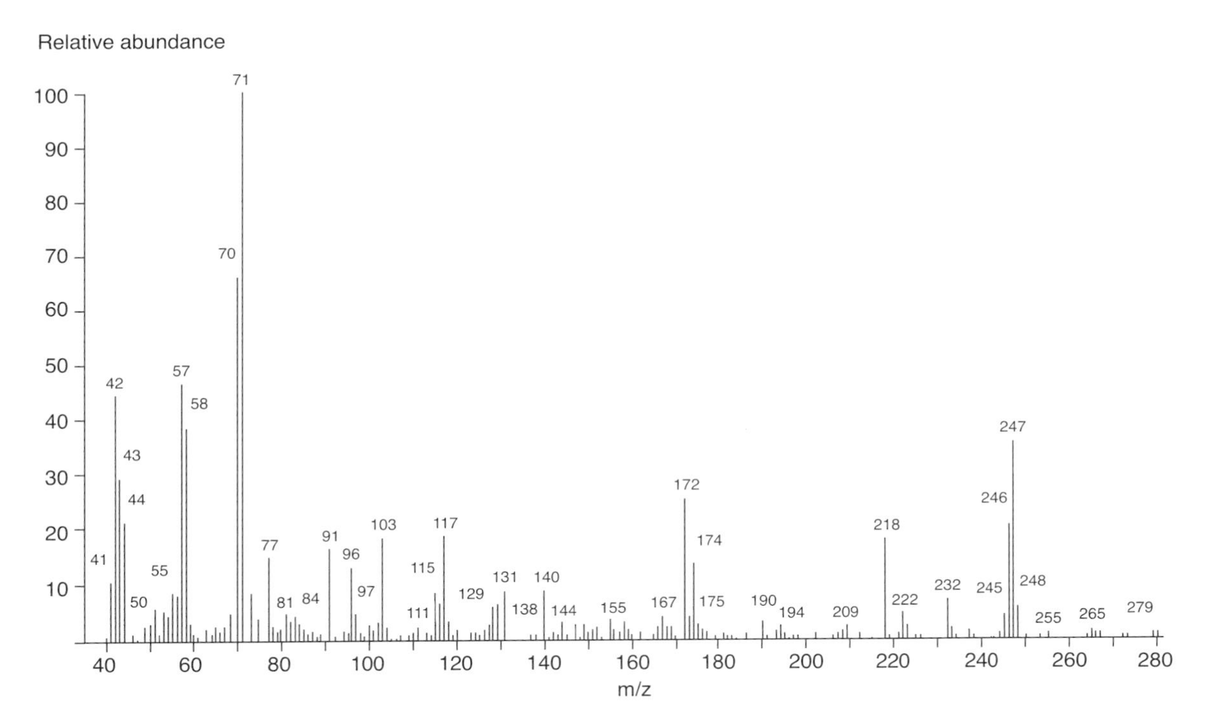

Figure 5 A typical mass spectrum

> **Q6. Explain why the inside of the mass spectrometer is kept under a high vacuum.**

Each vertical line in the mass spectrum represents a fragment of the original sample molecule. The line of highest mass (this is the peak furthest to the right, not the tallest peak) usually represents the unfragmented sample. The fragmentation process is not random; the same sample will always fragment in the same way provided the instrument conditions are kept the same. More stable fragments will tend to be more abundant and therefore produce taller peaks. So the mass spectrum can be used like a fingerprint to identify a substance by matching the spectrum of a suspect substance with that of a known sample of the same substance.

> **Q7. Figure 6 shows the mass spectrum of a known sample of pethidine. Compare it with Figure 5 which shows the mass spectrum of the peak in the GC with retention time 9.14 minutes. Do you think that the two represent a match? What features in the two spectra are similar and what features are different?**

RS•C

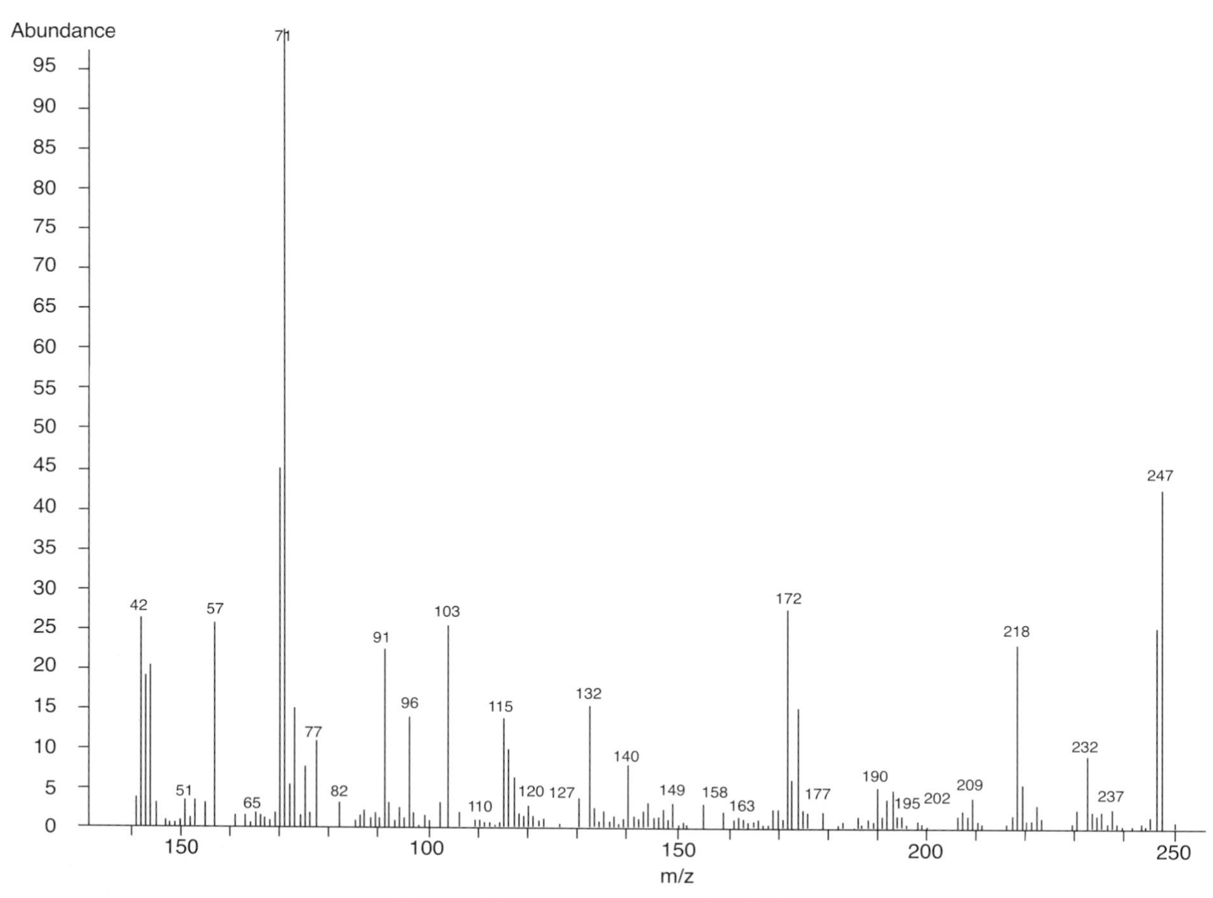

Figure 6 The mass spectrum of pethidine

At one time, spectra were compared visually as you will have done in the question above. Nowadays, computer software is available to search through a library of many thousands of mass spectra to find a match. It does this by comparing both masses at which peaks appear and the heights of the peaks.

Handling the data

GCMS generates an enormous amount of data – a chromatogram with maybe hundreds of peaks each of which produces a mass spectrum with many lines. All this data is stored and can be manipulated by a computer. One useful way of displaying this data is as a three-dimensional graph, see Figure 7. Here, the x-axis (horizontal) represents the retention time in the GC, the y-axis (vertically) represents amount and the z-axis (into the paper) represents the mass of the fragments.

RS•C

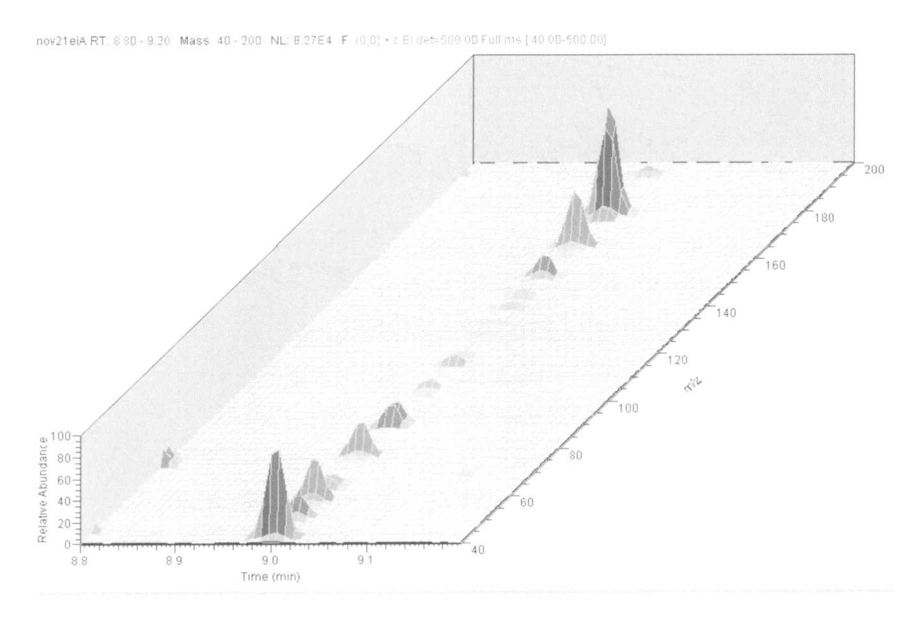

Figure 7

The data in Figure 7 can be stored, manipulated and displayed by a computer. For example the computer can present 'slices' through the three-dimensional array. A 'slice' in the plane of the paper represents a chromatogram while a slice at right angles to this (*ie* into the plane of the paper) represents the mass spectrum of one of the peaks.

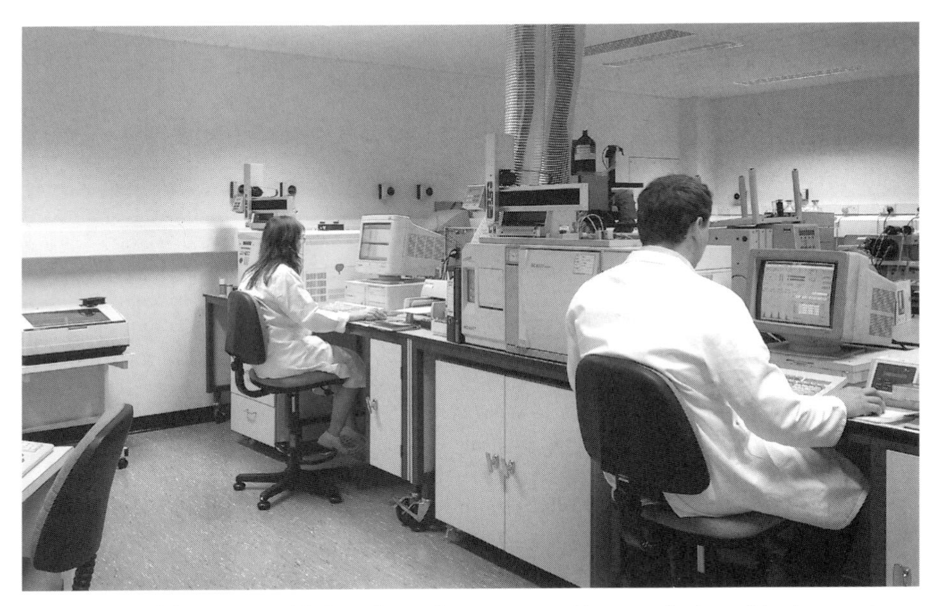

A GCMS laboratory with bench-top instruments. Data analysis and instrument control is carried out using PCs.

RS•C

Testing for drugs in racehorses using chromatography

Chromatography

You will probably have carried out chromatography, using filter paper and a liquid such as water or alcohol, to separate coloured dyes like those in ink or food colouring, see Figure 1. This is called paper chromatography and is one example of a whole range of methods. However they all work in basically the same way.

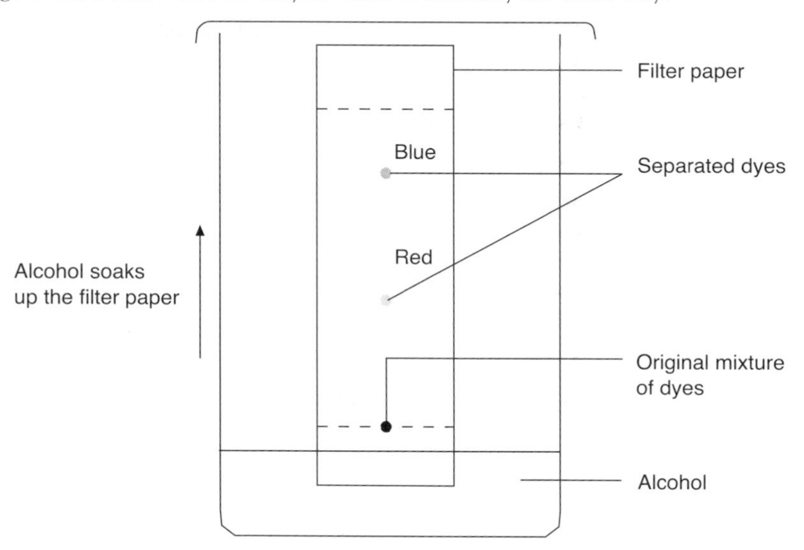

Figure 1 Paper chromatography using alcohol as the solvent

How chromatography works

All types of chromatography have a stationary phase (like the paper) and a mobile phase (such as alcohol). The stationary phase is usually a solid. The mobile phase may be either a liquid or a gas.

Mixtures separate as the mobile phase moves through the stationary phase (like water spreads through blotting paper) dragging the components (bits) of the mixture with it. Separation happens because some chemicals are more attracted to the paper than the alcohol and tend to lag behind. Others are more attracted to the alcohol than the paper and so move faster. So in Figure 1, the blue dye is more attracted to the alcohol than the red dye and so moves further in the same time.

Types of chromatography

Some of the types of chromatography are listed in Table 1.

Type	Stationary phase	Mobile phase
paper chromatography	paper	liquid
thin layer chromatography	alumina or silica powder on a glass plate	liquid
gas chromatography	solid or liquid coating a narrow tube	gas
liquid chromatography	solid	liquid
high performance liquid chromatography	solid	liquid

Table 1 Types of chromatography

Using chromatography to test for drugs in horses

Chromatography is used to detect drugs in horses (as well as in human athletes). Drugs may be used improperly because horseracing is a business as well as a sport. The owner of a winning horse gets prize money. People bet on races with bookmakers, and both want to make a profit. There is also a breeding industry, where the best male horses (stallions) and the best female horses (mares) are bred together. The owners of mares pay to have them bred with a stallion. The foals may be sold for large amounts of money.

Drug testing must be carried out for the following reasons:

- to prevent someone giving a horse a stimulant to improve its performance and help it win a race;
- to prevent criminals doping a horse to worsen its performance, so that bets on it are lost;
- to make sure that weakness is not introduced into breeding stock by making a horse seem better than it really is; and
- for animal welfare reasons – so the horses are not harmed by being given painkillers to allow them to run when they are injured, for example.

Samples taken from horses for drug testing are normally urine but may be blood. Urine samples are collected using a plastic bag held on a long pole. The sampler may rustle some straw – this trick tends to make horses urinate. Strict precautions have to be taken to ensure that the samples cannot be tampered with between being taken and arriving at the laboratory.

Q1. Suggest the kind of precautions that could be taken to prevent urine samples being tampered with between being taken and arriving at the laboratory.

RS•C

Sample kits are transported securely

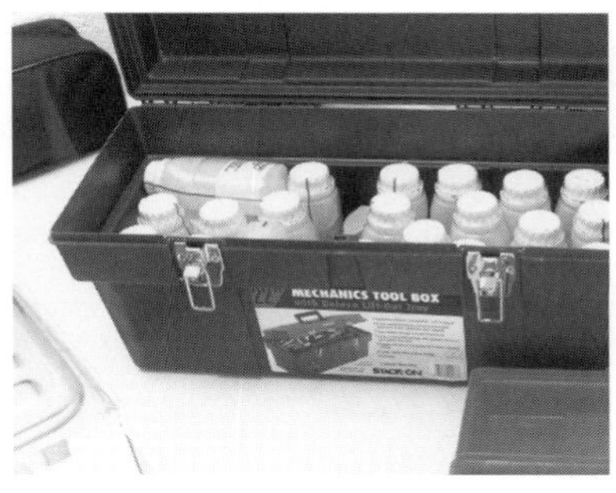

Inside the cases are a number of sampling bottles

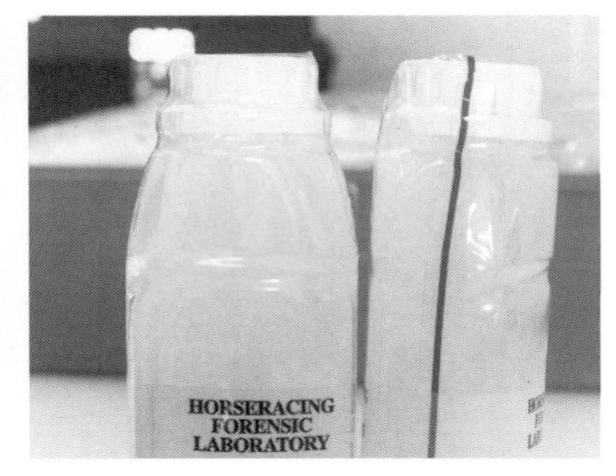

A new sample bottle is used for each sample.

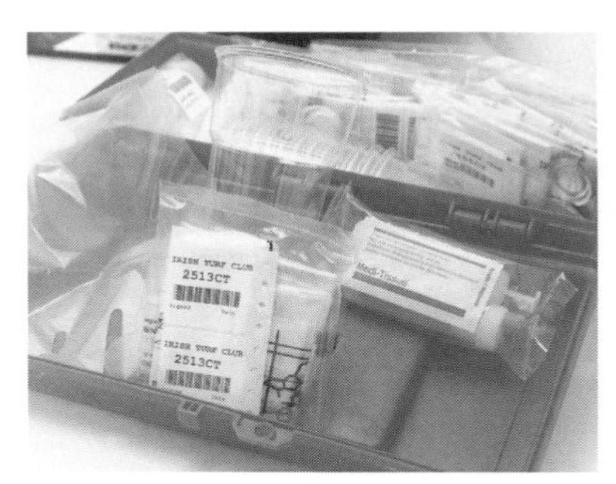

There are unique identifying labels and seals for each set of samples

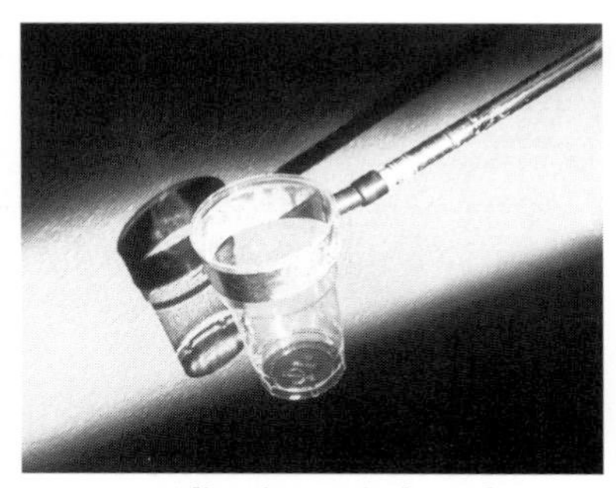
Sampling is carried out by persuading the horse to urinate into a beaker on a stick – not easy!

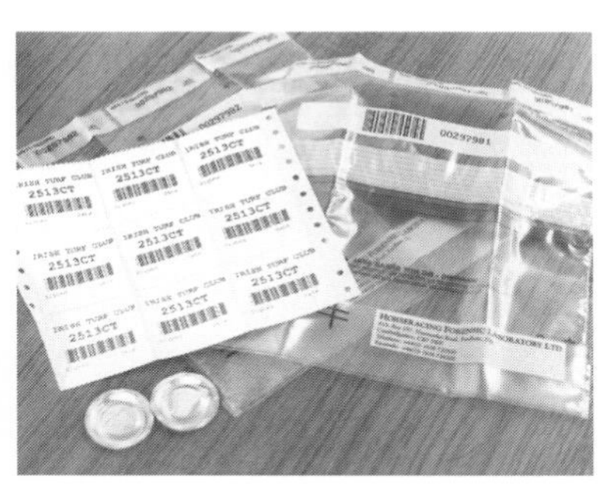

Bar codes are used to ensure the horse is not identified.

Sample bottles can be sealed using a thermal diaphragm

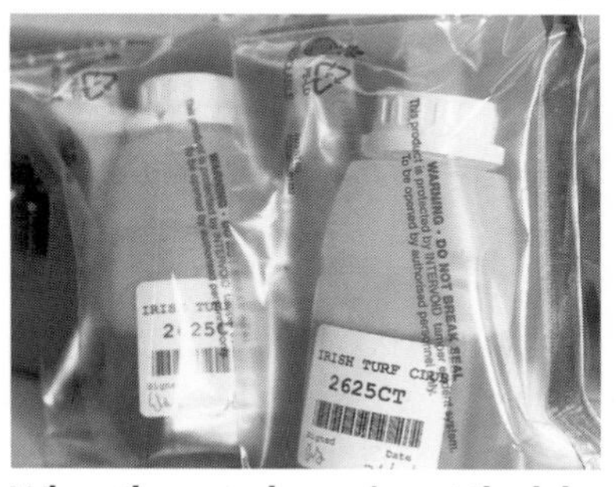

When the samples arrive at the lab, one is frozen and stored while the other is tested.

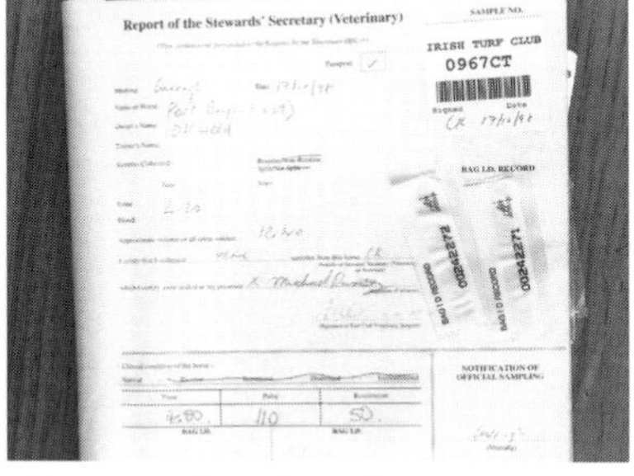

Paperwork and results are sent to the regulatory authority.

Figure 2 The sampling procedure

(Pictures reproduced by courtesy of the Irish Turf Club and Caroline Norris.)

In the UK, samples are divided into two portions and sent to the Horseracing Forensic Laboratory (HFL) near Newmarket in Suffolk. On arrival at the laboratory, one portion is kept frozen.

> **Q2. Suggest why the sample is divided into two and one half analysed and the other frozen.**

Sample preparation

The urine is first poured into a tube that has a tap at the lower end. This tube is filled with a white powder called silica. This setup is called a column (see Figure 3). The urine is

RS•C

washed through the column with two solutions of different pH. The first is alkaline and the second acidic. Many of the normal products contained in urine stay in the column. The rest of the material (including any drugs) is separated into two parts – those that are acid or neutral and those that are alkaline. Acidic and neutral compounds come out in the alkaline solution and alkaline components in the acidic one.

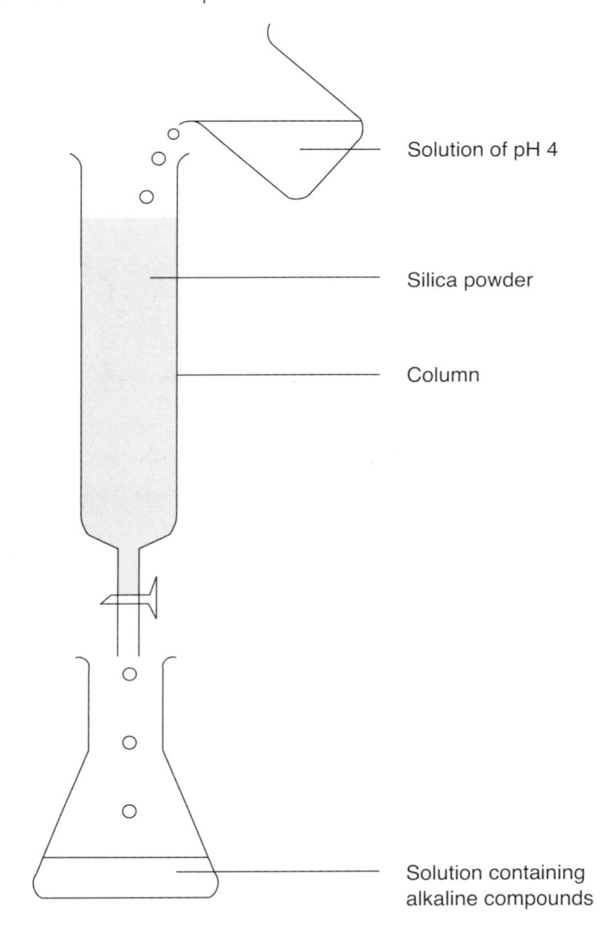

Figure 3 Separating drugs using a column

Q3. Look again at Figure 3 that shows the column that the urine is first separated on.

a) i) What is the stationary phase?
ii) What is the mobile phase?

b) How could you tell if the liquid in the conical flask was acidic, alkaline or neutral?

These parts are then separated by gas chromatography. In gas chromatography the stationary phase is a long, narrow tube made of silica (a material like glass) with a coating on the inside. This tube is also called a column. The mobile phase is a gas that flows through the tube, see Figure 4.

RS•C

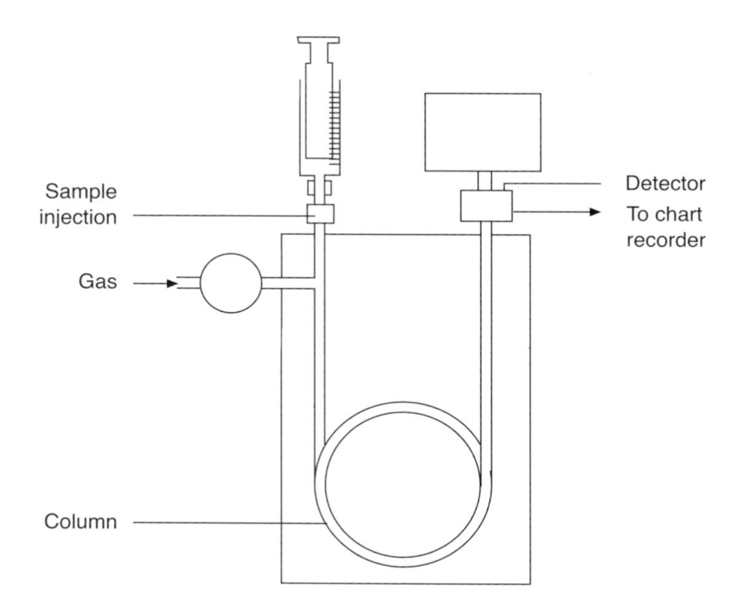

Figure 4 (a) The principle of the gas chromatography apparatus

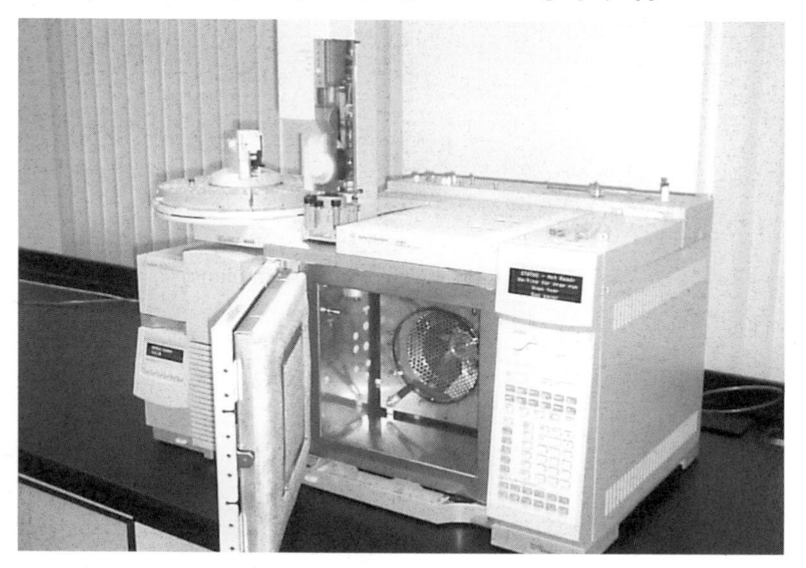

Figure 4 (b) Agilent 5973N GCMSD instrument, showing column inside oven

The separated materials come out of the tube at different times. As they are not coloured, we have to use an instrument to detect them. The results are plotted on a graph like Figure 5. Each spike on the graph is a different component of the urine. Any drug present will show up as an unusual spike.

RS•C

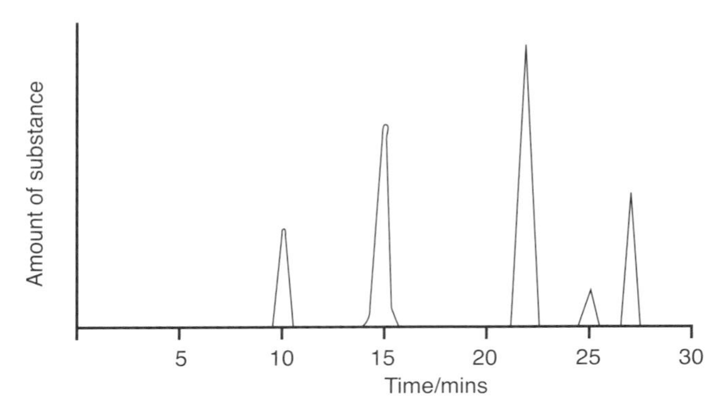

Figure 5 A gas chromatogram

> **Q4. In gas chromatography the columns are sometimes heated to over 200 ºC.**
>
> **a) Suggest why oxygen is not used to carry the substances through the column.**
>
> **b) Suggest a gas that could be used and explain your answer.**

One type of instrument used to detect the separated materials is called a mass spectrometer. The mass spectrometer can also help to identify these materials.

To be absolutely certain that the results are correct, the experiment is repeated using four samples:

1. the original urine;
2. water, to make sure that any chemicals used in the analysis are not contaminated (this is called the system blank)
3. urine known to be drug free (this is called the biological blank); and
4. a reference sample that contains known drugs to check that the analysis method detects these drugs.

If everything is as it should be, samples 2 and 3 will show no evidence of drugs.

If any drugs are found, the whole process is then checked by another chemist before a report is made to the Jockey Club, the body that controls horseracing in the UK.

> **Q5. An analysis was carried out and the results in Figure 6 obtained from the gas chromatogram.**
>
> **Was there a drug in the sample? Explain your answer.**

RS•C

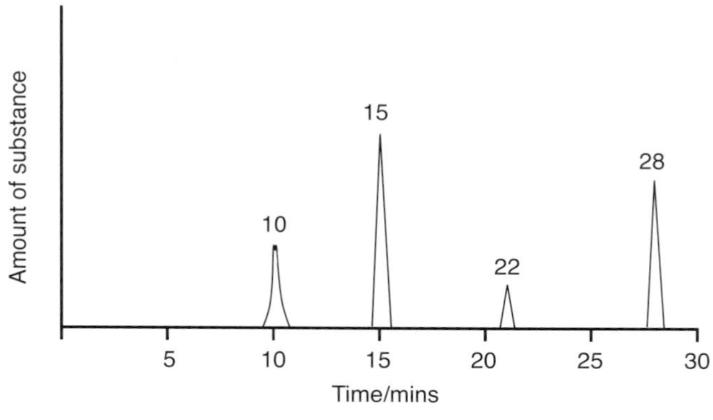

Figure 6a Sample

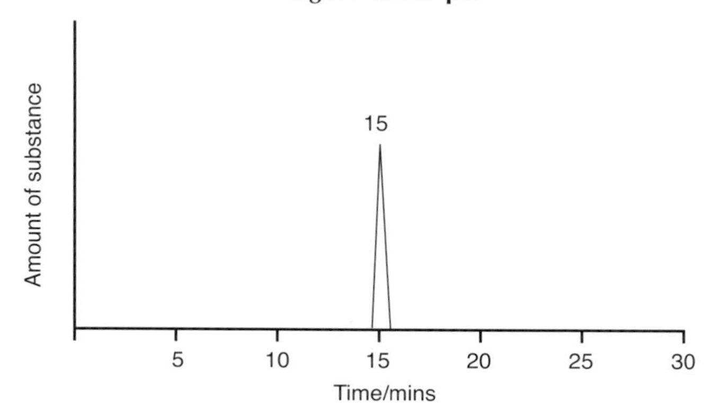

Figure 6b System blank

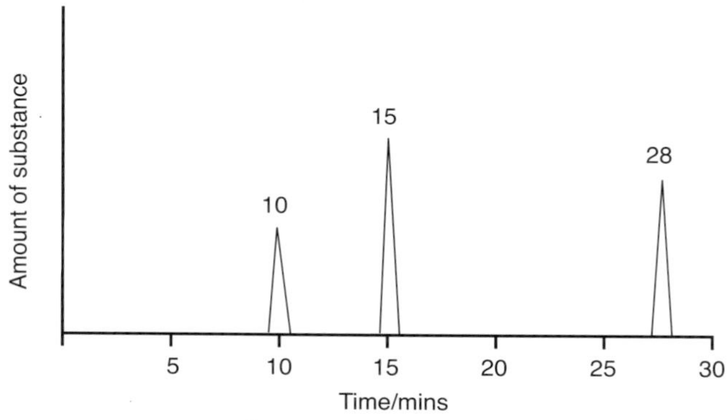

Figure 6c Biological blank

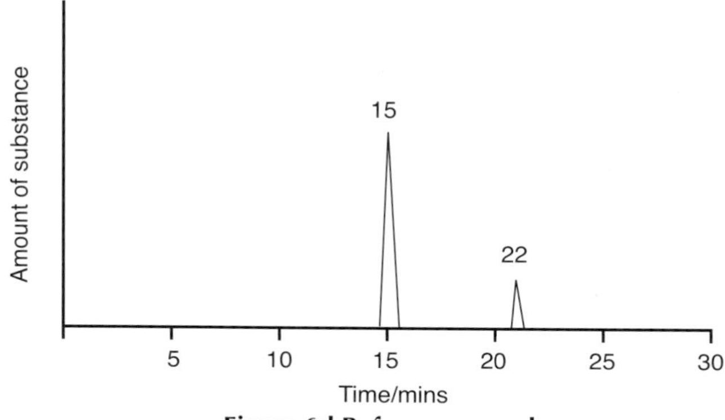

Figure 6d Reference sample

RS•C

What can be found

Often, if a drug is given to a horse, the drug itself is not found in the horse's urine. This is because chemical reactions in the horse's liver may change the drug. This is called metabolising the drug and the products are called metabolites. If the HFL chemists know what metabolites are found, they can work out what the original drug must have been.

Q6. Suggest how the HFL could prove that a particular compound is a metabolite of a drug that could be given to a horse. Explain in full the steps that would have to be carried out.

Q7. What other sports do you think could use this type of drug testing?

Q8. Human athletes sometimes take performance-enhancing drugs. What other reasons are there, apart from those in the passage, to test animal athletes?